NEW ENGLAND INSTITUTE
OF TECHNOLOGY
LEARNING RESOURCES CENTER

Handbook of
ELECTRIC MOTORS:
Use and Repair

Handbook of
ELECTRIC MOTORS:
Use and Repair

Second Edition

JOHN E. TRAISTER

NEW ENGLAND INSTITUTE
OF TECHNOLOGY
LEARNING RESOURCES CENTER

Published by
THE FAIRMONT PRESS, INC.
700 Indian Trail
Lilburn, GA 30247

Library of Congress Cataloging-in-Publication Data

Traister, John E.
   Handbook of electric motors : use and repair / by John E. Traister.
--2nd ed.
      p. cm.
   Includes index.
   ISBN 0-88173-144-7
   1. Electric motors. 2. Electric motors--Maintenance and repair.
I. Title.
TK2514.T73   1992   621.46--dc20   92-6857
                                                  CIP

Handbook Of Electric Motors: Use And Repair / By John E. Traister,
Second Edition

Copyright © 1992 by
PennWell Publishing Company
1421 South Sheridan / P.O. Box 1260 - Tulsa, Oklahoma 74104

All rights reserved. No part of this book may be reproduced, stored in a
retrieval system, or transcribed in any form or by any means, electronic or
mechanical, including photocopying and recording, without prior written
permission of the publisher.

Printed in the United States of America

10 9 8 7 6 5 4 3 2

ISBN 0-88173-144-7   FP

ISBN 0-13-379298-6   PH

While every effort is made to provide dependable information, the publisher, authors, and
editors cannot be held responsible for any errors or omissions.

Distributed by Prentice-Hall, Inc.
A Simon & Schuster Company
Englewood Cliffs, NJ 07632

Prentice-Hall International (UK) Limited, London
Prentice-Hall of Australia Pty. Limited, Sydney
Prentice-Hall Canada Inc., Toronto
Prentice-Hall Hispanoamericana, S.A., Mexico
Prentice-Hall of India Private Limited, New Delhi
Prentice-Hall of Japan, Inc., Tokyo
Simon & Schuster Asia Pte. Ltd., Singapore
Editora Prentice-Hall do Brasil, Ltda., Rio de Janeiro

Contents

Preface

Everyone associated in any capacity with the electrical industry will deal to some degree with electric motors and generators, as the electric motor is the principal means of changing electrical energy into mechanical energy to power all types of appliances and machines.

Most publications of the past have dealt quite heavily on theory and underlying principles, while brushing very lightly over the essential applications of electric motors. This publication does not slight the importance of certain principles that make possible the operation of rotary machinery, but more emphasis is placed on practical application of electric motors – the selection, use, and repair of them.

This book is not designed to be a "crash course" on electrical motor repair. Rather, it is meant to cover the entire field of motor applications, from the various available motor types to their use and repair. It cannot take the place of experience, but it can give the reader a very sound foundation upon which to build. The book should also prove useful to the experienced electrician or engineer as a review of motor applications.

Industrial electric shops will want to keep a copy of this book handy at all times for daily reference, and design and engineering departments will find constant use for the book in selecting motors for various applications, including machine design.

Over the past decade or so, many motor users have stopped doing their own repairs and have all work done by a qualified motor-repair shop. However, many of these same plants are now reconsidering the organization of their own in-house motor-repair shops to ensure prompt repair of motors that operate critical machines. This book should be extremely important to plants and personnel in this category.

A deep and grateful bow must be made in the direction of all manufacturers and others in the electrical industry who supplied helpful reference material and photos. This book could not have been completed without them. Credit lines appear throughout the book in their appropriate places, and names and addresses also appear in the appendix should the reader wish to contact them directly for further information on their products.

I am also indebted to Ros Herion of Prentice-Hall, who did the production work on this book. Her gentle (and sometimes hard) prodding kept the book moving along so that it could be published on time . . . and my wounds have just about healed by now. (Editor's note: Although a minimum of tongue-lashing was necessary, no whiplashing was required or employed.)

JOHN E. TRAISTER

Handbook of
ELECTRIC MOTORS:
Use and Repair

1

Introduction
to Electric Motors

The continuing rapid growth in the use of electricity for power and lighting is constantly spurred by a steady increase in the number and types of appliances, machines, and other devices powered by electric motors. In fact, the principal means of changing electrical energy into mechanical energy or power is the electric motor – ranging in size from small fractional horsepower, low-voltage motors to the very large high-voltage synchronous motors.

Electric motors are classified according to the following: size (horsepower); type of application; electrical characteristics; speed, starting, speed control, and torque characteristics; and mechanical protection characteristics and method of cooling.

Motor-control equipment varies from simple, thermal tumbler switches to large switchgear-type controllers. The type of controller used depends basically on the motor to be controlled, the type of control circuit, and the location of the controller. Usually the controller is mounted adjacent to the motor or in the immediate vicinity. In other instances, it may be at some distance away or may be incorporated in a motor control center at some distance remote from the motor it controls. Motor control centers may consist of a group of individual motor controllers connected by means of metal gutters and gutter connectors or a custom-built unit with all controls contained therein.

In basic terms, electric motors convert electrical energy into the productive power of rotary mechanical force. This capability finds application in unlimited ways from explosion-proof, water-cooled motors for underground mining to induced-draft fan motors for power generation; from adjustable-frequency drives for waste and water treatment pumping to eddy current clutches for automobile production; from direct-current drive systems for paper production to photographic film manufacturing; from rolled-shell shaftless motors for machine tools to large outdoor motors for crude oil pipelines; and from mechanical variable-speed drives for woodworking machines to complex adjustable-speed-drive systems for textiles.

All these and more represent the scope of electric motor participation in powering and controlling the machines and processes of industries throughout the world.

MAGNETISM

Anyone associated with the use and repair of electric motors should be familiar with the principles of magnetism because motors, as well as transformers, generators, and other electrical apparatus, depend on magnetism for their operation.

A magnet is either permanent or temporary. If a piece of iron or steel is magnetized and retains its magnetism, it is a permanent magnet. A compass, for example, is one form of permanent magnet. Others with which you are probably familiar are horseshoe-shaped magnets and bar magnets. Each one of these magnets has a north magnetic pole and a south magnetic pole; in fact, all magnets have a north and a south pole.

When current flows through a coil, a magnetic field with a north and a south pole is set up just like that of a permanent magnet. However, when the current stops, the magnetic field also disappears. This type of temporary magnetism is called electromagnetism. Permanent magnets are used for the magnetic field necessary in the operation of small, inexpensive electrical motors. Larger motors, relays, and transformers rely upon the magnetic fields from electrical current passing through coils.

When electricity flows through a wire or conductor, magnetic lines of force (magnetic flux) are created around the wire as shown in Fig. 1-1.

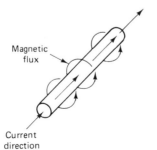

Magnetic flux

Current direction

FIG. 1-1 **When current flows through a conductor, magnetic flux is created around the wire.**

When a piece of wire is passed through a magnetic field (magnetic lines of force), electricity is created in that wire. We then can readily see the relation between electricity and magnetism. In fact, the very existence of the electrical industry is dependent on magnetism and magnetic circuits.

ELECTROMAGNETIC INDUCTION

Whenever a wire or other conductor is moved in a magnetic field so that the conductor cuts across the lines of force, there is an electromotive force produced in that conductor. If a conductor remains stationary and the magnetic field moves so that its lines cut across the conductor, an electromotive force is produced; "inducing" electromotive force by movement between a conductor and a magnetic field is called electromagnetic induction.

If a piece of wire is moved through magnetic lines of force (see Fig. 1-2) so the wire cuts across the path of the flux, a voltage will be induced in this wire. If a sensitive meter is then connected to this wire – completing the circuit – the dial will indicate a flow of current every time the wire is moved across the lines of force.

Referring again to Fig. 1-2, if the wire is moved upward through the flux, the meter will read to the left of zero, assuming the zero mark is in the center of the scale. If the wire is moved downward through the flux, the needle or indicator will read to the right of zero. Should the wire be moved rapidly up and down through the flux, the needle will swing back and forth – to the left and right of the zero mark. From this experiment it can be seen that the direction of the induced voltage and resulting current flow depends on the direction of movement through the magnetic field and that both voltage and current can be reversed merely by reversing the direction of movement of the wire. If the wire is held in the magnetic flux with no movement, no voltage will be generated. Or if the wire

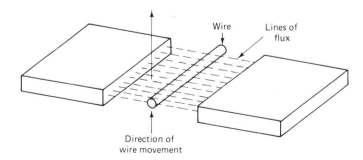

FIG. 1-2   Moving a wire through magnetic flux in such a way that it cuts across the path of the flux induces a voltage in the wire.

is moved to the right or left – parallel to the path of the flux – no voltage will be produced. The wire must cut across the flux path to generate voltage – in other words, cutting the lines of force.

The faster the wire is moved through the magnetic field or the stronger the field and greater the number of lines of force, the more voltage will be generated. Therefore, the amount of voltage produced by electromagnetic induction depends on the speed with which lines of force are cut or the number of lines cut in a given period of time.

For example, one conductor cutting 100 million (100,000,000) lines of force per second will produce 1 V of pressure. This obviously is an enormous number of lines to cut to produce 1 V of electricity. However, by speeding up the movement of the conductor, as in an actual generator, the conductor will pass many poles per second, causing a higher voltage. The voltage may also be increased by connecting several wires in series in the form of coils.

For example, three separate wires moved upward through the flux at the same time causes an equal amount of voltage to be induced in each, all in the same direction. Then, when all are connected in series so their voltages will all add up in the same direction in the circuit, there will be three times as much voltage as with one wire. Generator coils are often made with many hundreds of turns connected in series to produce a very high voltage.

The direction of the induced voltage depends on the direction of the flux and on the direction in which the conductor is moved. One way to determine the direction of induced voltage when the directions of the motion and of the flux are known is to apply Fleming's right-hand rule:

> If the thumb of the right hand is pointed in the direction of the motion and the forefinger is pointed in the direction of the flux, the middle finger will be pointing in the direction of the generated voltage.

The position of the right hand when the conductor is moved upward is shown in Fig. 1-3. When the conductor is moved downward, the position of the hand must be changed so that the thumb points downward; in this case, the middle finger will point forward, indicating a reversed voltage.

An elementary alternator is shown in Fig. 1-4. A stationary permanent magnet 1, called a field magnet, with the poles N and S (north and south) provides the magnetic field. A rectangular coil of wire 2 is connected to the slip rings 3 and 4, and both the coil and the rings are mounted on a shaft, not shown, which is caused to rotate by some outside source of mechanical power. Stationary pieces of carbon, called brushes, 5 and 6, bear against the slip rings and provide connections to the load 7 in the external circuit.

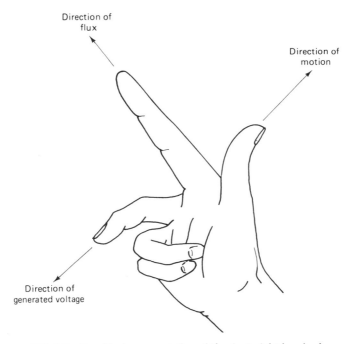

Direction of
flux

Direction of
motion

Direction of
generated voltage

FIG. 1-3   Graphical representation of Fleming's right-hand rule.

When the coil is rotated in either direction, a voltage is induced in this coil, but this voltage changes direction every half revolution.

Assume that the coil in Fig. 1-4 is rotating counterclockwise between the poles as indicated by the curved arrow. The lower side of the coil is moving upward toward the pole center, and so, according to Fleming's right-hand rule, the induced voltage is in the direction of the arrow, or toward the back of the machine. At the same time, the upper side of the coil is moving downward, also cutting the lines of force, and the

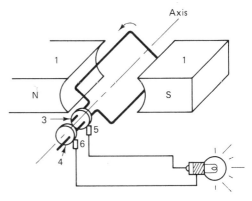

Axis

N

S

1

1

3

5

6

4

FIG. 1-4   An elementary alternator consists of a field magnet, a coil of wire, and slip rings. Carbon brushes bear against the slip rings to provide connections to the load in the external circuit.

voltage induced in this side is toward the front, as indicated by the arrow. When the coil is connected through the slip rings and brushes to the outside load, a current starts to flow in the circuit in the direction indicated by the arrows: toward the back in the lower side of the coil, around the coil back, through the upper coil side toward the front, through slip ring 3 and brush 5 to the load, and then to brush 6 and slip ring 4 and back to the coil. At one particular instant, the voltages induced in the two sides of the coil have such directions that the current flows in the same direction in all parts of the circuit.

As the coil approaches the center of the poles, the induced voltage increases because more lines of force are cut. The maximum voltage is reached as the coil sides pass the horizontal position. As the rotation continues, the coil sides cut fewer lines of force, and the induced voltage decreases until it reaches zero value when the coil is in an upright position and does not cut any lines of force. As the coil continues to rotate, the direction of the voltage in each coil side reverses, because the side which moved upward before now moves downward, and the side which moved downward before now moves upward. The current through the load circuit now flows in the direction opposite to that indicated by the arrows in Fig. 1-4. The voltage first reaches a maximum in value and then decreases to zero when the coil is in an upright position between the poles. If the coil continues to rotate, the same process is repeated for every revolution.

The changes of induced voltage during one revolution are called one cycle. In each half of the cycle the voltage changes its direction and is, therefore, an alternating voltage, or ac voltage. If the machine is built so that there are 60 cycles during 1 s, it is said that the voltage and the current have a frequency of 60 cps.

POWER AND ENERGY

Energy, in brief, means ability to do work. Mechanical work is done when any kind of energy is used to produce motion in a body formerly stationary, or to increase the rate of motion of a body, or to slow down its rate of motion.

The most generally used unit of work is the foot-pound, which is when a mass of 1 lb is lifted 1 ft against the force of gravity. The total amount of work done is equal to the number of feet of motion multiplied by the number of pounds moved. Time doesn't enter into the matter of mechanical work.

MECHANICAL POWER.   Power is the rate of doing work involving work and time. One of the units in which power may be measured is *foot-pounds per second.* Power is equal to the total amount of work divided by the time taken to do the work. It is assumed that the work is being done at a uniform or constant rate, at least during the period of time measured.

The foot-pound per second is a unit too small to be used in practice. Mechanical power most often is measured in the unit called a *horsepower.* One horsepower is the power rate corresponding to 550 ft·lb/s. Since there are 60 s in a minute, 1 hp also corresponds to 550 times 60, or to 33,000 ft·lb/min.

An electric power at the rate of 1 hp would be capable of raising a weight of 33,000 lb through a distance of 1 ft in 1 min. At the same rate of 1 hp the motor would lift during 1 min any number of pounds through a distance such that the pounds times the number of feet equaled 33,000.

ELECTRIC POWER.   For an electric motor to continue working at the rate of 1 hp, an electric current must be sent through the motor at a certain number of amperes when the pressure difference across the motor terminals is a certain number of volts. The number of amperes and the number of volts would have to be such that multiplied together they would equal 746. A unit of electric power to describe this product of amperes and volts is called the watt. One watt is the power produced by a current of 1 A when the pressure difference is 1 V.

The total number of watts of power is equal to the number of amperes of current multiplied by the number of volts of pressure difference, both with reference to the device in which power is being produced. The number of volts must be 746 to produce 1 hp, so 746 W of electric power is equivalent to 1 mechanical hp.

One watt-hour of electric energy is the quantity of energy used with a power rate of 1 W when this rate continues for 1 h. That is, the watt-hours of energy are equal to the number of watts multiplied by the number of hours during which power is used at this rate. A 60-W electric lamp uses power at the rate of 60 W so long as it is lighted to normal brilliancy. But the total quantity of energy used by the lamp depends also on the total length of time it remains lighted. If the 60-W lamp is kept lighted for 10 h, it will have used 60 × 10, or 600 W·h of electric energy.

BASIC ELECTRIC MOTOR PRINCIPLES

Electric motors are machines that change electrical energy into mechanical energy. They are rated in horsepower. The attraction and repulsion of the magnetic poles produced by sending current through the armature and field windings cause the armature to rotate. The armature rotation produces a twisting power called torque.

FLEMING'S LEFT-HAND RULE FOR MOTORS. Place the thumb, first finger, and remaining fingers at right angles to each other. Point the first finger in the direction of the field flux and the remaining fingers in the direction of the armature current, and the thumb will indicate the direction of rotation.

The direction of rotation can be reversed on any dc motor by reversing either the armature or field leads but not both. It is standard practice to reverse the armature leads to reverse the direction of rotation.

The amount of torque developed by a motor is proportional to the strength of the armature and field poles. Increasing the current in the armature or field winding will increase the torque of any motor.

The armature conductors rotating through the field flux have a voltage generated in them that opposes the applied voltage. This opposing voltage is called counterelectromotive force (CEMF) and serves as a governor for the dc motor. After a motor attains normal speed, the current through the armature will be governed by the CEMF generated in the armature winding. This value will always be in proportion to the mechanical load on the motor.

The applied voltage is the line voltage. The effective voltage is the voltage used to force the current through the resistance of the armature winding. This value can be determined by multiplying the resistance of the armature by the current flow through it. To find the resistance of the armature, measure the voltage drop across the armature and the current flow through it and use the Ohm's law formula: R equals E over I.

The hp rating of a motor refers to the rate of doing work. The amount of hp output is proportional to the speed and torque developed by the motor.

BASIC MOTOR TERMS AND NAMEPLATE INFORMATION

Before covering electric motor applications and details of their maintenance and repair, the reader should be thoroughly familiar with certain motor terms. Some of the more basic ones follow and are provided courtesy of Westinghouse Electric Corp.

Style number: Identifies that particular Westinghouse motor from any other. Other manufacturers also provide style numbers on the motor nameplate and in the written specifications.

Serial data code: The first letter is a manufacturing code used at the factory. The second letter identifies the month, and the last two numbers identify the year of manufacture (D78 is April 78).

Frame (fr):   Specifies the shaft height and motor mounting dimensions and provides recommendations for standard shaft diameters and usable shaft extension lengths.

Service factor: A service factor is a multiplier which, when applied to the rated horsepower, indicates a permissible horsepower loading which may be carried continuously when the voltage and frequency are maintained at the value specified on the nameplate, although the motor will operate at an increased temperature rise.

NEMA service factors: Open motors only.

hp	S.F.	hp	S.F.	hp	S.F.
$\frac{1}{12}$	1.40	$\frac{1}{3}$	1.35	1	1.15*
$\frac{1}{8}$	1.40	$\frac{1}{2}$	1.25	$1\frac{1}{2}$	1.15
$\frac{1}{6}$	1.35	$\frac{3}{4}$	1.25	2	1.15
$\frac{1}{4}$	1.35			3	1.15

*hp, open motor at 3450 rpm, has 1.25 service factor.

Phase: Indicates whether the motor has been designed for single or three phase. It is determined by the electrical power source.

Thermoguard® motors: Incorporates an overload protector, which is a heat-sensing device either attached to the motor winding or mounted in the end bell. The two most common types of Thermoguard motors are the following:

1. *Type A:* This is an automatic thermal device approved by UL. It will stop the motor when it overheats and will automatically restart the motor when it has cooled to a safe operating temperature.

2. *Type M:* This is a manual reset overload device which is approved by UL. It will also stop the motor when it overheats but will not start unless the thermal protector button is manually reset.

Degree C ambient: The air temperature immediately surrounding the motor. Forty degrees centigrade is the NEMA maximum ambient temperature.

Insulation class: The insulation system is chosen to ensure the motor will perform at the rated horsepower and service factor load.

Horsepower: Defines the rated output capacity of the motor. It is based on breakdown torque, which is the maximum torque a motor will develop without an abrupt drop in speed.

rpm: Revolutions per minute . . . speed. The rpm reading on motors is the approximate full-load speed.

The speed of the motor is determined by the number of poles in the winding.

A two-pole motor runs at an approximate speed of 3450 rpm. A four-pole motor runs at an approximate speed of 1725 rpm. A six-pole motor runs at an approximate speed of 1140 rpm.

Amps: Gives the amperes of current the motor draws at full load. When two values are shown on the nameplate, the motor usually has a dual voltage rating. Volts and amps are inversely proportional; the higher the voltage, the lower the amperes, and vice versa. The higher amp value corresponds to the lower voltage rating on the nameplate. Two-speed motors will also show two ampere readings.

Hertz (cycles per second): Just about everything in this country is serviced by 60-Hz alternating current. Therefore, most applications will be for 60-Hz operations.

Volts: Volts is the electrical potential "pressure" for which the motor is designed. Sometimes two voltages are listed on the nameplate, such as 115/230. In this case the motor is intended for use on either a 115 or 230 circuit. Special instructions are furnished for connecting the motor for each of the different voltages.

kVA code: This code letter is defined by NEMA standards to designate the locked rotor kVA per horsepower of a motor. It relates to starting current and selection of fuse or circuit breaker size.

Housing: Designates the type of motor enclosure. The most common types are open and enclosed: *Open drip-proof* has ventilating openings so constructed that successful operation is not interfered with when drops of liquid or solid particles strike or enter the enclosure at any angle from 0° to 15° downward from the vertical. *Open guarded* has all openings giving direct access to live metal or hazardous rotating parts so sized or shielded as to prevent accidental contact as defined by probes illustrated in the NEMA standard. *Totally enclosed* motors are so constructed to prevent the free exchange of air between the inside and outside of the motor casing. *Totally enclosed fan-cooled* motors are equipped for external cooling by means of a fan that is integral with the motor. *Air-over* motors must be mounted in the airstream to obtain their nameplate rating without overheating. An air-over motor may be either open or enclosed.

Explosion-proof motors: These are totally enclosed designs built to withstand an explosion of gas or vapor within the motor and to prevent ignition of the gas or vapor surrounding the motor by sparks or explosions which may occur within the motor casing.

Hours: Designates the duty cycle of a motor. Most fractional horsepower motors are marked continuous for around-the-clock operation at the nameplate rating in the rated ambient. Motors marked "one half" are for ½-h ratings, and those marked "one" are for 1-h ratings.

The following sections consist of terms not found on the nameplate, but they are important considerations for proper motor selection.

BEARINGS TERMS

Sleeve bearings: Sleeve bearings are generally recommended for axial thrust loads of 210 lb or less and are designed to operate in any mounting position as long as the belt pull is not against the bearing window. On light-duty applications, sleeve bearings can be expected to perform a minimum of 25,000 h without relubrication.

Ball bearings: These are recommended where axial thrust exceeds 20 lb. They too can be mounted in any position. Standard and general-purpose ball bearing motors are factory-lubricated and under normal conditions will require no additional lubrication for many years.

MOUNTING TERMS

Rigid mounting: A rectangular steel mounting plate which is welded to the motor frame or cast integral with the frame; the most common type of mounting.

Resilient mounting: A mounting base which is isolated from motor vibration by means of rubber rings secured to the end bells.

Flange mounting: A special end bell with a machined flange which has two or more holes through which bolts are secured. Flange mountings are commonly used on such applications as jet pumps and oil burners.

Rotation: For single-phase motors, the standard rotation, unless otherwise noted, is counterclockwise facing the lead or opposite shaft end. All motors can be reconnected at the terminal board for opposite rotation unless otherwise indicated.

2

Split-Phase Motors

Split-phase motors are fractional hp units which use an auxiliary winding on the stator to aid in starting the motor until it reaches its proper rotation speed. See Fig. 2-1. This type of motor finds uses in small pumps, oil burners, automatic washers, and other household appliances.

In general, the split-phase motor consists of a housing, a laminated iron core stator with embedded windings forming the inside of the cylindrical housing, a rotor which is made up of copper bars set in slots in an iron core and connected to each other by copper rings around both ends of the core and plates which are bolted to the housing and contain the bearings which support the rotor shaft, and a centrifugal switch inside the housing. This type of rotor is often called a *squirrel cage* rotor since the configuration of the copper bars resembles an actual cage. Such motors have no windings as such, and a centrifugal switch is provided to open the circuit to the starting winding when the motor reaches running speed.

To understand the operation of a split-phase motor, look at the wiring diagram in Fig. 2-1. Current is applied to the stator windings, both the main winding and the starting winding which is in parallel with it, through the centrifugal switch. The two windings set up a rotating magnetic field, and this field sets up a voltage in the copper bars of the squirrel cage rotor. Because these bars are shortened at the ends of the rotor, current flows through the rotor bars. The current-carrying rotor bars then react with the magnetic field to produce motor action. When the rotor is turning at the proper speed, the centrifugal switch cuts out the starting winding since it is no longer needed.

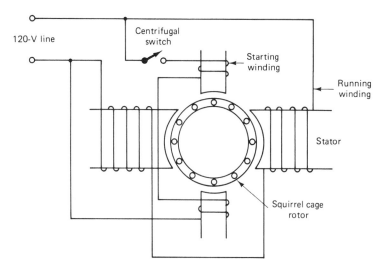

FIG. 2-1   Schematic diagram of a standard split-phase ac motor.

CENTRIFUGAL SWITCH

The centrifugal switch, located inside the motor, consists of two main parts: a stationary part and a rotating part. The former is located on the front end plate of the motor and has two contacts, similar to a conventional single-pole switch. The latter, or rotating part, is located on the rotor as shown in Fig. 2-2.

When the motor is not running, the two contacts on the stationary part of the centrifugal switch are kept closed by the pressure of the rotating part. However, at about 75% of the motor's full speed, the rotating part releases its pressure against the contacts and causes them to open, disconnecting the starting winding from the circuit.

The stationary part of some centrifugal switches consists of two copper semicircular segments that are insulated from each other and mounted on the inside of the front end plate. The rotating part has three

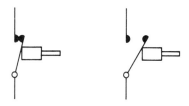

copper fingers that ride around the stationary segments while the motor is starting. While the motor is at a standstill, the segments are shortened by the copper fingers, causing the starting winding to be included in the motor circuit. When the motor reaches approximately 75% of its full speed, the centrifugal force causes the fingers to be lifted from the segments, disconnecting the starting winding from the circuit.

SQUIRREL-CAGE ROTOR

The rotors of induction motors are either wound or of the squirrel cage type. The type used on split-phase motors is of the latter type. It consists of conducting bars connected rigidly together by conducting rings. The voltage induced in each bar is seldom more than 10 V, so that insulation between the bars and core is not always necessary. The bars under each pole are in parallel with the currents under adjacent poles in opposite directions. The current flowing out from the bars under one pole passes into the end ring and, dividing equally, returns to the other side of the core through the bars under each of the two adjacent half poles as shown in Fig. 2-3.

Since the squirrel-cage rotor is symmetrical and has no connections within itself – which determine the relative directions of current flow in the rotor – the currents induced in the bars and the number of poles produced are determined only by the stator winding. This enables a squirrel-cage rotor to operate in any primary of the correct mechanical dimensions and will run at the speed of the stator field minus the slip. However, the starting torque, which is dependent on the rotor resistance, will not be the same when the rotor is used in stators with different numbers of poles, since the current paths in the rotor winding are changed, altering the effective resistance of the secondary.

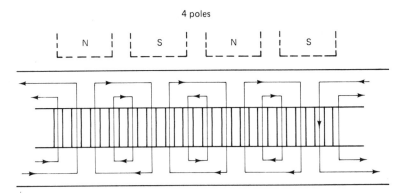

FIG. 2-3   Current paths in a squirrel cage rotor.

The rotor resistance is made up of bar and end-ring resistance. Since the bars under each pole are in parallel, the bar resistance per short-circuit path will be less as the number of poles is decreased and the number of bars per pole increases. But the number of parallel paths decreases as the bars per pole increase. Since these two effects will then balance each other, the equivalent bar resistance does not change as the number of poles is varied.

Changing the number of primary poles, however, changes the effective resistance of the portion of the winding which is included in the end rings. The cross-sectional area of the ring is constant, while the length of the ring included by each path decreases and the number of parallel paths increases as the number of poles is increased. Therefore, the portion of the rotor resistance due to the end rings is approximately inversely proportional to the square of the number of poles.

In general, the greater the resistance of a squirrel-cage winding, the greater is the starting torque and the slower the speed corresponding to any torque. High-resistance windings for slow-speed, multipolar machines are made by using bars of small cross section or of low conductivity and rings made of a low-conductivity alloy.

In actual practice, the size and number of bars and the size of the rings are usually standardized, so that the required total resistance is obtained by choosing different combinations of available components. Therefore, no definite percentage of the total resistance is put in the bar or ring. For this reason, one motor can have its secondary resistance, and consequently its starting torque, increased by machining the end rings, while another motor may show but little increase in torque after machining off a large portion of the rings.

SPLIT-PHASE MOTOR TROUBLES

When a split-phase motor fails to operate properly, a definite procedure should be followed in determining the repairs necessary to put it into good condition to operate properly. To do so means performing a series of tests to discover the exact trouble. These tests should quickly show if the motor needs only minor repairs, such as new bearings, or if a partial or complete overhaul is in order. See Fig. 2-4.

FAILURE TO START

1. *No voltage:* Use a voltmeter to check for proper voltage at the motor terminals. If the readings indicate ZERO voltage, check voltage, check for blown fuses or a tripped circuit breaker.

2. *Low voltage:* Take a reading at the motor terminals with the

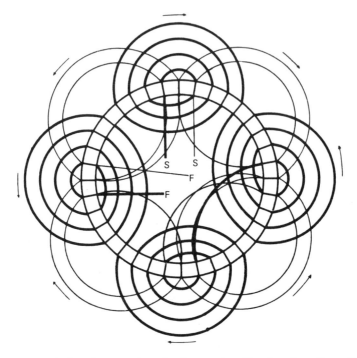

FIG. 2-4   Winding diagram of a split-phase motor. (Courtesy Page Power Co.)

switch closed. This voltage should be within 10% of the rated nan plate voltage. Overloaded transformers or circuits can cause a lc voltage condition.

3. *Faulty centrifugal switch:* Remove the inspection plate in 1 front end bracket to observe the switch operation as the circui1 energized. If the disc does not hold the switch in the closed posit when the motor is standing still, the motor cannot start; the ε play washers may need adjusting. Check the contact points for and foreign matter. Also check for clogged weights and a w spring.

4. *Open overload-protection device:* If the motor is equipped · a built-in overload-protection device, let the motor cool before ing to start it. If this fails, remove the cover plate on the end bra on which the switch is mounted and see if the switch contact: closed. If the switch is stuck, remove and repair.

5. *Grounded field:* If the motor overheats or produces a sl when touched or the output power at idle is excessive, test f grounded field by placing the leads of a continuity tester betv the field leads and the frame. If the field is grounded, repai replace the motor.

6. *Open field:* Apply a current to both the starting and running windings separately through a test-lamp circuit. However, do not leave the windings connected too long while the rotor is stationary. If either winding is open, repair or replace the motor.

7. *Shorted field:* If the motor draws excessive current and at the same time lacks sufficient torque, overheats, or hums, a shorted field is indicated. The motor must be repaired or replaced.

8. *Faulty end play:* Some split-phase motors have steel-enclosed cork washers at each end of the motor to cushion the end thrust. Excessive end thrust, heat, or hammering on the shaft can destroy the washers and interfere with the operation of the cutout-switch mechanism. If necessary, install new end-thrust cushion-bumper assemblies. The end play should be adjusted so that the cutout switch is closed at standstill and open when the motor is operating.

9. *Excessive load:* This may be determined by comparing the input current with the current rating on the nameplate. Excessive load may prevent the motor from reaching the speed at which the governor acts, thereby causing the phase winding to become overheated and burn out.

10. *Tight bearings:* Turn the armature by hand to determine if the bearings are too tight. If they are stiff and tight, add an oil that has been specified by the manufacturer. If this procedure does not work, replace the bearings.

LESS THAN SATISFACTORY OPERATION

1. *Motor runs hot:* Check for a grounded or shorted field, tight bearing, faulty centrifugal switch, wring voltage, or excessive load.

2. *Motor does not reach full speed:* Perform the same check as in test 1.

3. *Excessive bearing wear:* Check belt tension and alignment. Check for dirty, incorrect, or insufficient oil. If bearings are clogged with dirt or other foreign matter, clean them thoroughly. Worn bearings should be replaced.

4. *Excessive noise:* Check for worn bearings and excessive end play. If necessary, add additional end-play washers. Check for loose motor parts, hold-down bolts, or pulley; bad alignment, worn belts, sprung shafts, or unbalanced rotor; and/or burrs on the shaft shoulders.

5. *Motor produces shock:* Check for grounded stator; cushion-mounted motors have a ground strip that carries static electricity across the rubber mounting to ground. If the static charge is retained, the strip may be broken, or there may be poor connections present. The frames of all motors should be grounded.

6. *Rotor rubs stator:* Clean burrs and dirt from both the rotor and the stator. Check for worn bearings.

7. *Radio and TV interference:* Check for poor ground connections. Static electricity generated by the motor belts may cause noises if the motor is not thoroughly grounded. Check for loose contacts in the switch, the fuses, and/or the starter.

In performing the preceding tests, the following steps may prove to be useful and save considerable time:

- Carefully examine the outside of the motor housing to detect any broken or cracked parts – such as broken end plates or a bent shaft. Remove the cover plate and inspect for burned motor leads.
- Check for bearing trouble by moving the shaft up and down in the bearing. Any movement in either direction indicates a worn bearing. Turn the rotor by hand and observe the freedom of movement. A shaft that does not rotate freely is an indication of severe problems – either bad bearings, a bent shaft, or an improperly assembled motor. If any of these problems exists, the motor should run hot and trip the overload protective devices.
- Examine the connection box carefully to see if any bare leads (either motor or line) are touching the motor frame. Also check to see if the enternal wires are making contact with the iron cores of the rotor or stator. Use a test meter here.
- If all the preceding tests check out favorably, connect the power lines to the motor terminals and see if the motor runs. If there are internal problems, usually the overcurrent device will trip or blow, the windings may smoke, or the motor may rotate slowly or noisily. Furthermore, the rotor may not turn at all; rather, humming may be heard. Such symptoms always indicate internal problems. If problems are evident, the motor should be disassembled and tested more carefully as described elsewhere in this book.

3

Capacitor Motors

Capacitor motors are single-phase ac motors ranging in size from fractional hp to perhaps as high as 15 hp. This type of motor is widely used in all types of single-phase applications such as powering machine shop tools (lathes, drill presses, etc.), air compressors, refrigerators, and the like. This type of motor is similar in construction to the split-phase motor (Chapter 2), except a capacitor is wired in series with the starting winding as shown in Fig. 3.1.

The capacitor provides higher starting torque, with lower starting current, than the split-phase motor, and although the capacitor is sometimes mounted inside the motor housing, it is more often mounted on top of the motor — encased in a metal compartment.

In general, there are two types of capacitor motors in use: the capacitor-start motor and the capacitor-start-and-run motor. As the name implies, the former utilizes the capacitor only for starting; it is disconnected from the circuit once the motor reaches running speed, or about 75% of the motor's full speed. Then the centrifugal switch opens to cut the capacitor out of the circuit.

The capacitor-start-and-run motor keeps the capacitor and starting winding in parallel with the running winding at all times, providing quiet and smooth operation at all times.

As a rule, single-phase motors are used in sizes up to about 7½ hp, and the capacitor-start motors are utilized for heavy-duty loads in residential and small commercial applications for which three-phase service is not practical or not available at all.

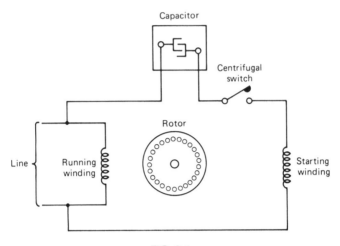

FIG. 3-1

The capacitor-start motor is available in normal starting torque de-signs for such loads as fans, blowers, centrifugal pumps, and similar applications. The high-starting torque designs are used on such equip-ment as reciprocating compressors, pumps, loaded conveyors, and the like.

Capacitor split-phase motors require the least maintenance of all single-phase motors, but they have a very low starting torque, making them unsuitable for many applications. Their high maximum torque, however, makes them especially useful for such tools as floor sanders or in grinders where momentary overloads due to excessive cutting pressure are experienced. They are also used quite frequently for slow-speed direct-connected fans.

Refer to the diagram in Fig. 3-1; both the starting and running wind-ings are connected across the line during the starting period since the centrifugal switch is closed. Note that the starting winding is connected in series with the capacitor and the centrifugal switch and remains in this position during the starting cycle. Then, when the motor reaches approx-imately 75% of its full speed, the centrifugal switch opens, cutting out both the capacitor and starting winding from the line circuit.

The starting torque in a capacitor motor is produced by a revolving magnetic field which is established inside the motor. This magnetic field is accomplished by locating the starting winding 90 electrical deg out of phase with the running winding, causing the current in the starting wind-ing to lead the current in the running winding. This condition produces a revolving magnetic field in the stator, which in turn induces a current in the rotor winding, producing rotation of the motor.

CAPACITORS

Physically, a capacitor consists of two conducting materials separated by an insulator. Its purpose is to oppose any change in voltage, and it has the ability to store electrical energy. When connected to an ac power supply – as in a capacitor-start motor – the plates will charge alternately positive and negative and give the illusion of current flow through the insulating medium. Because of the charge-discharge cycle, the current must lead its voltage by 90 electrical deg enabling its power factor to be zero and consuming no power.

Capacitors are rated in microfarads (μ F), and the type most often used in motors is the electrolyte capacitor. This type of capacitor consists of two sheets of aluminum foil which are separated by one or more layers of gauze which have been saturated with a chemical solution called an electrolyte. See Fig. 3-2. This type of capacitor is suitable only for intermittent duty and should not be held in a circuit for more than a few seconds at a time.

In checking capacitors, remove them from the circuit and then measure capacitance on a capacitance bridge if one is available. This test will detect either open or shorted capacitors. Measure the resistance of a capacitor with an ohmmeter. If the capacitor is shorted, the meter will indicate less than 10 Ω. If the capacitor is good, the resistance will be 50 MΩ or greater.

To check for an open-circuited capacitor, connect as shown in Fig. 3-3. During this test, if the neon tester does not glow, the capacitor is defective. This test, however, does not apply to mica grid capacitors. Resistance between terminals, and between terminals and case, measured with a megohmmeter, should be at least 50 MΩ. In the case of mica grid capacitors, check either on a capacitance bridge or try a new capacitor. For a short-circuit test of a mica grid capacitor, leave the capacitor on the

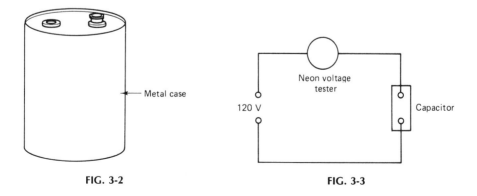

FIG. 3-2 FIG. 3-3

tube socket, remove the external grid lead, and measure the resistance
with an ohmmeter. It should show at least 100 MΩ.

The main reasons for a capacitor becoming defective include the following: excessive use, overheating, and held in the circuit too long. In the
case of capacitors for capacitor-start motors, a defective capacitor must
be replaced with one of approximately the original value of capacitance,
or the motor may not have the required starting torque for proper starting and operation.

4

Repulsion-Type Motors

Repulsion-type motors are divided into several groups including repulsion-start, induction-run motors; repulsion motors; and repulsion-induction motors. Regardless of the varying types, there are certain construction characteristics which are common to all of them:

- Each has a stator similar to the running winding on a split-phase motor.
- The rotor consists of a slotted core with embedded windings connected to a commutator at one end of the rotor.
- Bearings are mounted in the end plates to support the rotor shaft.
- Carbon brushes are fitted in holders and ride on the commutator to provide a path for current flow between commutator segments and each brush.
- Brush holders are supported in the motor, either on the front end plate or on the rotor shaft.

REPULSION-START, INDUCTION-RUN MOTORS

The repulsion-start, induction-run type of motor is of the single-phase type, ranging in size from about $\frac{1}{10}$ hp to as high as 20 hp. It has high starting torque and a constant-speed characteristic, which makes it suitable for such applications as commercial refrigerators, compressors, pumps, and similar applications requiring high starting torque.

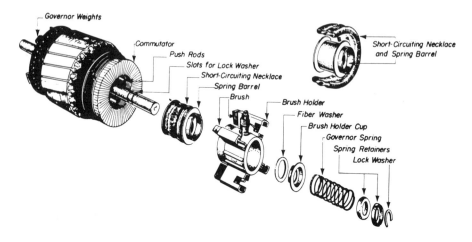

FIG. 4-1

There are two different designs of repulsion-start, induction-run motors: the brush-lifting type and the brush-riding type. In the former, the brushes are automatically moved away from the commutator when the motor reaches approximately 75% of its full speed. In the brush-riding type, the brushes ride on the commutator at all times. The brush-riding arrangement (with an axial form of commutator) is used almost exclusively on smaller motors, whereas the brush-lifting type (with a radial form of commutator) is used in both small and large motors. See Fig. 4-1.

REPULSION MOTORS

The repulsion motor is distinguished from the repulsion-start, induction-run motor by the fact that it is made exclusively as a brush-riding type and does not have any centrifugal mechanism. Therefore, this motor both starts and runs on the repulsion principle. This type of motor has high starting torque and a variable-speed characteristic. It is reversed by shifting the brush holder to either side of the neutral position. Its speed can be decreased by moving the brush holder farther away from the neutral position.

The stator of the repulsion motor is like that of the repulsion-start, induction-run motor, and the stator is generally wound for four, six, or eight poles. Four leads are normally brought out of the motor for dual voltage operation.

The rotor consists of an armature constructed much in the same manner as those used in dc motors with laminations and is also generally skewed. The windings may be either hand or coil wound and are connected either with lap or wave windings. The commutator is of the axial type, and the brushes always ride on the commutator.

REPULSION-START-INDUCTION MOTORS

In the repulsion-start-induction type of motor are combined the high starting torque of the repulsion type and the good speed regulation of the induction motor. The stator of this motor is provided with a regular single-phase winding, while the rotor winding is similar to that used on a dc motor. When starting, the changing single-phase stator flux cuts across the rotor windings, inducing currents in them; when flowing through the commutator, a continuous repulsive action upon the stator poles is present.

This motor starts as a straight repulsion type and accelerates to about 75% of normal full speed when a centrifugally operated device connects all the commutator bars together and converts the winding to an equivalent squirrel cage type. The same mechanism usually raises the brushes to reduce noise and wear. Note that when the machine is operating as a repulsion type the rotor and stator poles reverse at the same instant and that the current in the commutator and brushes is ac.

This type of motor will develop four to five times normal full-load torque and will draw about three times normal full-load current when starting with full-line voltage applied. The speed variation from no load to full load will not exceed 5% of normal full-load speed.

The repulsion-start-induction motor is used to power air compressors, refrigeration, pumps, meat grinders, small lathes, small conveyors, stokers, and the like. In general, this type of motor is suitable for any load that requires a high starting torque and constant-speed operation. Most motors of this type are less than 5 hp.

Troubles occurring in this type of motor can be found in the commutator, brushes, centrifugal switch, short-circuited rig, bearings, oil-soaked insulation, solder thrown out of commutator, too much or too little tension on the throw-out centrifugal spring, and opens, shorts, or grounds in the rotor of stator windings. Rotation is reversed by shifting the brushes.

MOTOR DRIVES

The type of motor selected for a given application should be of sufficient capacity to avoid overloading but not so large to cause it to be inefficient and to make the initial cost higher than it should be. It is usually better to install a small motor on each machine being driven than to install one large motor to operate a drive shaft to operate all the machines.

The proper design of motor drives for machines involves the selection of a method of connecting the motor to the driven machine in an adequate and efficient manner. The selection depends on power supply, space limitations, safety and working conditions for operators, initial cost and operating costs, production and quality of product, and surrounding conditions. The objective is to obtain a working installation which will be the best when all these items are considered.

Motors can be connected to machines in several ways. When the speed of the motor is identical to the speed desired for the driven machine, a direct coupling is used to connect the motor shaft with the machine shaft. When the speed of the motor is different from the speed of the machine shaft, a connection must be used which will give the proper speed ratio between the motor and the machine. A speed ratio may be obtained by using a V belt, flat belt, chain or set of gears. With belts, the speed is inversely proportional to the effective diameter of the two pulleys. With a chain or gears, the speed ratio is inversely proportional to the number of teeth on the two sprockets or gears.

5

Polyphase Motors

Three-phase motors offer extremely efficient and economical application and are usually the preferred type for commercial and industrial applications when three-phase service is available. These motors are available in ratings from fractional hp up to thousands of hp in practically every standard voltage and frequency. In fact, there are few applications in which the three-phase motor cannot be put to use.

Three-phase motors are noted for their relatively constant-speed characteristic and are available in designs giving a variety of torque characteristics; that is, some have a high starting torque while others a low starting torque. Some are designed to draw a normal starting current and others, a high starting current.

A typical three-phase motor is shown in Fig. 5-1. Note that the three main parts are the stator, rotor, and end plates. It is very similar in construction to conventional split-phase motors except that the three-phase motor has no centrifugal switch.

The stator shown in Fig. 5-2 consists of a steel frame and a laminated iron core and a winding formed of individual coils placed in slots. The rotor may be a squirrel cage or wound rotor type. Both types contain a laminated core pressed onto a shaft. The squirrel cage rotor is shown in Fig. 5-3 and is similar to a split-phase motor. The wound rotor is shown in Fig. 5-4 and has a winding on the core that is connected to slip rings mounted on the shaft.

The end plates or brackets are bolted to each side of the stator frame and contain the bearings in which the shaft revolves. Either ball bearings or sleeve bearings are used.

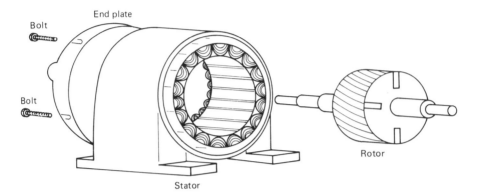

FIG. 5-1 Typical three-phase motor construction and features. (Courtesy Marathon Electric.)

FIG. 5-2 Stator of a three-phase motor showing the coil assembly and iron core separately. (Courtesy Dreisilker Electric Motors.)

FIG. 5-3   Typical squirrel cage rotor.
(Courtesy Westinghouse.)

FIG. 5-4   Phase-wound, or wound, rotor. (Courtesy Dreisilker Electric Motors.)

OPERATING PRINCIPLES OF TWO-PHASE MOTORS

Two-phase motors are designed to operate on two-phase alternating current and have two windings, each covering one half of each pole, or spaced 90° apart, similar to the starting and running windings of a single-phase motor.

Each of the windings in a two-phase motor, however, is of the same size wire and has the same number of turns. Instead of being wound with spiral coils, two-phase windings are generally made with diamond-shaped coils similar to those used in armatures. A section of a two-phase winding is shown in Fig. 5-5. Note the manner in which the three coils of each phase overlap in forming the winding for one pole of the motor.

In Fig. 5-6 are shown the curves for two-phase current with alternations 90° apart. When the current flows through the two windings, it sets up poles that progress step by step around the stator so rapidly that it produces what is practically a revolving magnetic field. The progress of this field and the magnetic poles can be observed by tracing out and comparing the illustration in Fig. 5-7 to the one in Fig. 5-6. The dashed lines running vertically through the curves in Fig. 5-6 indicate the polarity of the curves at that instant and are referred to as *positions*.

In position 1 of Fig. 5-6, *A* and *B* are both positive, and, referring to

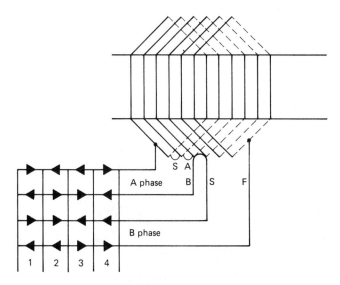

FIG. 5-5   Instantaneous current direction of coil groups.

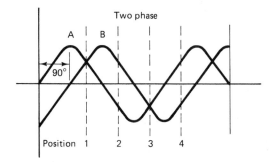

FIG. 5-6   Diagram of two-phase current.

Position 4	+	+	+	·	·	·	·	·	·	+	+	+
Position 3	·	·	·	·	·	·	+	+	+	+	+	+
Position 2	·	·	·	+	+	+	+	+		·	·	·
Position 1	(+)	(+)	(+)	(+)	(+)	(+)	(·)	(·)	(·)	(·)	(·)	(·)

In   Out

FIG. 5-7   Rotation of field in two-phase motor.

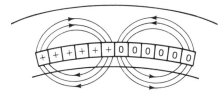

FIG. 5-8   Position of magnetic flux in position 1.

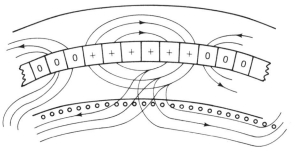

FIG. 5-9   Position of magnetic flux in position 2.

position 1 at the leads of the windings in Fig. 5-5, current will flow in the starting leads of the two windings marked S and S. The polarity set up will be as shown by the positive and negative marks in Fig. 5-7 at position 1. At this instant, the current flows in at all six wires on the left (Fig. 5-5) and out at all six wires on the right. See Fig. 5-7, bottom line. This will set up a magnetic flux or polarity as shown in the sketch of the magnetic circuit, position 1, shown in Fig. 5-8, showing that the center of the pole at this instant will be in the exact center of the coils and that a north pole will be produced at this point on the stator teeth.

At position 2 in Fig. 5-6, the B phase is still positive, but the A phase has changed to negative. Therefore, the current in the starting lead of the A phase will reverse as shown at position 2 and cause a reversal of polarity around the A group. As this group covers the first half of the pole, these three slots will change in polarity. The first three slots of the second pole will also change and cause the pole to move three slots to the right as shown in position 2 in Fig. 5-7. This shift of the magnetic pole is also illustrated in the magnetic sketch in Fig. 5-9.

At position 3 in Fig. 5-6, *B* has changed to negative, and the current in the leads of the *B* phase (Fig. 5-5) will reverse, causing the last three slots in each pole to change in polarity so the center of the pole moves three more slots to the right, as shown in Fig. 5-7.

The currents in the coil groups reverse in the manner just described and keep shifting the magnetic poles to the right; a corresponding change or movement of the field takes place in the stator. As this flux moves to the right and cuts across the rotor bars, it induces currents in them, and the reaction between the poles of this secondary current in the rotor and the stator poles causes the field of the stator poles to be distorted from its natural shape. It is from this distortion that the torque or twisting force is produced and causes the rotor to turn.

OPERATING PRINCIPLES OF THREE-PHASE MOTORS

The rotating action of the field in a three-phase motor is very much the same as that of two-phase motors, with the exception that only one third of the pole, or two slots, reverses at a time. In the two-phase motor, one half of the pole, or three slots, changes at each reversal of current. The coil groups of the three-phase winding should be placed in the slots in such a manner that they alternate in the same order as the currents change in the three-phase system.

Observe the three-phase current curves in Fig. 5-10 and note that

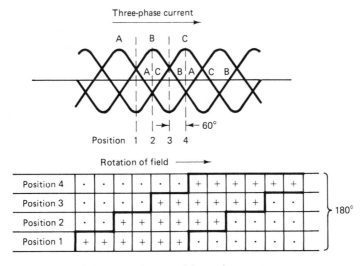

FIG. 5-10   Diagram of three-phase current.

the alternations change polarity or cross the center line in the order *A*, *C*, *B*, etc. The coil groups should be wound in to correspond with these current changes, or in the order *A*, *C*, *B*, etc.

At any given time in three-phase systems, the voltage or current curves above the zero line will exactly equal those below the line as can be seen in Fig. 5-10.

There is another condition that always exists in three-phase windings. Trace the current in toward the winding on the line wires in Fig. 5-11 and note that the center group, or *C* phase, will be traced around the coils in the direction opposite to the *A* and *B* phases. Keep in mind that the three currents never flow toward the winding at the same time and that there will always be a return current on one of the wires. Therefore, at any time when all three wires are carrying current, there will either be two positives and one negative or two negatives and one positive.

When these three currents flow through a three-phase winding, three consecutive coil groups will be of the same polarity, and the next three groups will be of opposite polarity, building up alternate poles; that is, N.S., N.S., etc. From this, it should be seen how the field poles progress around the stator to produce a revolving magnetic field in a three-phase motor.

Trace out and compare each of the positions (1, 2, 3, and 4) in Fig. 5-10 and compare them to the diagram in Fig. 5-11; note the manner in which the field poles gradually advance in the slots as the current alternates in the three phases, *A*, *C*, and *B*.

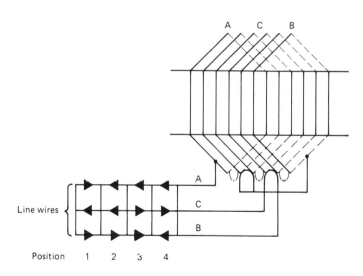

FIG. 5-11   Instantaneous current direction of three-phase motor.

POLYPHASE MOTOR ENCLOSURES

Polyphase induction motors differ in construction and appearance, depending on the type of service for which they are to be used. Open and closed frames are quite common. In the former enclosure, the motor's parts are covered for protection, but the air can freely enter the enclosure. Further designations for this type of enclosure include drip-proof, weather-protected, and splash-proof.

Totally enclosed motors, such as the one shown in Fig. 5-12, have an airtight enclosure. They may be fan-cooled or self-ventilated. An enclosed motor equipped with a fan (Fig. 5-12) has the fan as an integral part of the machine but external to the enclosed parts. In the self-ventilated enclosure, no external means of cooling is provided.

The type of enclosure used will depend on the ambient and surrounding conditions. In a drip-proof machine (see Fig. 5-13), for example, all ventilating openings are so constructed that drops of liquid or solid particles falling on the machine at an angle of not greater than 15° from the vertical cannot enter the machine, even directly or by striking and running along a horizontal or inclined surface of the machine. The application of this machine would lend itself to areas where liquids are processed.

An open motor having all air openings which give direct access to live or rotating parts, other than the shaft, limited in size by the design of the parts or by screen to prevent accidental contact with such parts is classified as a drip-proof, fully guarded machine. In such enclosures, openings shall not permit the passage of a cylindrical rod ½ in. in diameter, except where the distance from the guard to the live rotating parts is more than 4 in., in which case the openings shall not permit the passage of a cylindrical rod ¾ in. in diameter. See Fig. 5-14.

There are other special applications of drip-proof machines, such as externally ventilated and pipe ventilated, which, as the names imply, are

FIG. 5-12 Totally enclosed, fan-cooled motor. (Courtesy Leeson Electric Corporation.)

FIG. 5-13   Open, drip-proof motor. (Courtesy Leeson Electric Corporation.)

FIG. 5-14   Totally enclosed fan-cooled motor. (Courtesy Leeson Electric Corporation.)

either ventilated by a separate motor-driven blower or cooled by ventilating air from inlet ducts or pipes.

An enclosed motor whose enclosure is designed and constructed to withstand an explosion of a specified gas or vapor which may occur within the motor and to prevent the ignition of this gas or vapor surrounding the machine is designated an *explosion-proof (XP)* motor.

Hazardous atmospheres (requiring XP enclosures) of both a gaseous and dusty nature are classified by the NE Code as follows:

Class I, Group A: atmospheres containing acetylene
Class I, Group B: atmospheres containing hydrogen or gases or vapors of equivalent hazards such as manufactured gas

Class I, Group C: atmospheres containing ethylether vapor

Class I, Group D: atmospheres containing gasoline, petroleum, naphtha, alcohols, acetone, lacquer-solvent vapors, and natural gas

Class II, Group E: atmospheres containing metal dust

Class II, Group F: atmospheres containing carbon-black, coal, or coke dust

Class II, Group G: atmospheres containing grain dust

The proper motor enclosure must be selected to fit the particular atmospheres. However, explosion-proof equipment is not generally available for class I, groups A and B, and it is therefore necessary to isolate motors from the hazardous area.

MOTOR TYPE

The type of motor will determine the electrical characteristics of the design. NEMA has designated the following designs for polyphase motors:

NEMA Design	Starting Torque	Starting Current	Breakdown Torque	Full-Load Slip
A	Normal	Normal	High	Low
B	Normal	Low	Medium	Low
C	High	Low	Normal	Low
D	Very high	Low	–	High

An A motor is a three-phase, squirrel cage motor designed to withstand full voltage starting with locked rotor current higher than the values for a B motor and having a slip at rated load of less than 5%.

A B motor is a three-phase, squirrel cage motor designed to withstand full voltage starting and developing locked rotor and breakdown torques adequate for general application and having a slip at rating load of less than 5%.

A C motor is a three-phase, squirrel cage motor designed to withstand full voltage starting, developing locked rotor torque for special high-torque applications, and having a slip at rated load of less than 5%.

Design D is also a three-phase, squirrel cage motor designed to withstand full voltage starting, developing 275% locked rotor torque, and having a slip at rated load of 5% or more.

SYNCHRONOUS MOTORS

A synchronous polyphase motor has a stator constructed in the same way as the stator of a conventional induction motor. The iron core has slots into which coils are wound which are also arranged and connected in the same way as the stator coils of the induction motor. These are in turn grouped to form a three-phase connection, and the three free leads are connected to a three-phase source. Frames are equipped with air ducts which aid the cooling of the windings, and the coil guards protect the winding from damage.

The rotor of a synchronous motor carries poles which project toward the armature and are called *salient poles*. The coils are wound on laminated pole bodies and connected to slip rings on the shaft. A squirrel cage winding for starting the motor is embedded in the pole faces.

The pole coils are energized by direct current, which is usually supplied by a small dc generator called the *exciter*. This exciter may be mounted directly on the shaft to generate dc voltage which is applied through brushes to slip rings. On low-speed synchronous motors, the exciter is normally belted or of a separate high-speed motor-driven type.

The dimensions and construction of synchronous motors vary greatly, depending on the rating of the motors. However, synchronous motors for industrial power applications are rarely built for less than 25 hp or so. In fact, most are 100 hp or more. All are polyphase motors when built in this size. Vertical and horizontal shafts with various bearing arrangements and various enclosures cause wide variations in the appearance of the synchronous motor.

Synchronous motors are used in electrical systems where there is need for improvement in power factor or where a low power factor is not desirable. This type of motor is especially adapted to heavy loads that operate for long periods of time without stopping, such as for air compressors, pumps, ship propulsion, and the like.

The construction of the synchronous motor is well adapted for high voltages, as it permits good insulation. Synchronous motors are frequently used on 2300 V or more. Their efficient slow-running speed is another advantage.

NATIONAL ELECTRICAL CODE REQUIREMENTS

Article 430 of the National Electrical Code covers rules governing motors, motor circuits, and controllers, including phase converters. All such equipment must be installed in a location that allows adequate ventilation to cool the motors. Furthermore, the motors should be located so that maintenance, troubleshooting and repairs can be readily performed. Such work could consist of lubricating the motor's bearings, or perhaps replacing worn brushes..Testing the motor for open circuits and ground faults is also necessary from time to time.

When motors must be installed in locations where combustible material, dust, or similar material may be present, special precautions must be taken in selecting and installing motors.

Exposed live parts of motors operating at 50 volts or more between terminals must be guarded; that is, they must be installed in a room, enclosure, or location so as to allow access by only qualified persons (electrical maintenance personnel). If such a room, enclosure, or location is not feasible, an alternative is to elevate the motors not less than 8 feet above the floor. In all cases, adequate space must be provided around motors with exposed live parts—even when properly grounded—to allow for maintenance, troubleshooting, and repairs.

The chart in Figure 5-15 summarizes installation rules for the 1990 NE Code. Further, detailed information may be found in the NE Code book, under the Articles or Sections indicated in the chart.

Application	NE Code Regulation	NE Code Section
Location	Motors must be installed in areas with adequate ventilation. They must also be arranged so that sufficient work space is provided for replacement and maintenance.	430-14
	Open motors must be located or protected so that sparks cannot reach combustible materials.	430-16
	In locations where dust or flying material will collect on or in motors in such quantities as to seriously interfere with the ventilation or cooling of motors and thereby cause dangerous temperatures, suitable types of enclosed motors that will not overheat under the prevailing conditions must be used.	
Disconnecting means	A motor disconnecting means must be within sight from the controller location (with exceptions) and disconnect both the motor and controller. The disconnect must be readily accessible and clearly indicate the *Off/On* positions (open/closed).	Article 430 (H)

	Motor-control circuits require a disconnecting means to disconnect them from all supply sources.	430-74
	The service switch may serve to disconnect a single motor if it complies with other rules for a disconnecting means.	430-106
	The disconnecting means must be a motor-circuit safety switch rated in horsepower or a circuit breaker.	430-109
Wiring methods	Flexible connections such as Type AC cable, "Greenfield," flexible metallic tubing, etc. are standard for motor connections	Articles 310 and 430
Motor-control circuits	All conductors or a remote motor control circuit outside of the control device must be installed in a raceway or otherwise protected. The circuit must be wired so that an accidental ground in the control device will not start the motor.	430-73
Guards	Exposed live parts of motors and controllers operating at 50 volts or more must be guarded by installation in a room, enclosure, or location so as to allow access by only qualified persons, or elevated 8 feet or more above the floor.	Article 430
Motors operating over 600 volts	Special installation rules apply to motors operating at over 600 volts.	Article 30 (J)
Controller grounding	Motor controllers must have their enclosures grounded.	430-144

Figure 5-15 Summary or NE Code requirements for motor installations.

6

Alternating-Current Motor Control

Motor controllers cover a wide range of types and sizes from a simple toggle switch to a complex system with such components as relays, timers, and switches. The common function, however, is the same in all cases: to control some operation of an electric motor. An electric motor controller will include some of all of the following functions:

- Starting and stopping
- Overload protection
- Overcurrent protection
- Reversing
- Changing speed
- Jogging
- Plugging
- Sequence control
- Pilot light indication

The controller can also provide the control for auxiliary equipment such as brakes, clutches, solenoids, heaters, and signals and may be used to control a single motor or a group of motors.

The term *motor starter* is often used in the electrical industry and means practically the same thing as *controller*. Strictly speaking, a motor starter is the simplest form of controller and is capable of starting and stopping the motor and providing it with overload protection.

MANUAL STARTER

A manual starter is a motor controller whose contact mechanism is operated by a mechanical linkage from a toggle handle or push button which is in turn operated by hand. A thermal unit and direct-acting overload mechanism provides motor-running overload protection. Basically, a manual starter is an *on-off* switch with overload relays.

Manual starters are used mostly on small machine tools, fans and blowers, pumps, compressors, and conveyors. They cost the least of all motor starters, have a simple mechanism, and provide quiet operation with no ac magnet hum. The contacts, however, remain closed and the lever stays in the on position in the event of a power failure, causing the motor to automatically restart when the power returns. Therefore, low-voltage protection and low-voltage release are not possible with these manually operated starters. However, this action is an advantage when the starter is applied to motors that run continuously.

Fractional horsepower manual starters are designed to control and provide overload protection for motors of 1 hp or less on 120- or 240-V single-phase circuits. They are available in single- and two-pole versions and are operated by a toggle handle on the front. When a serious overload occurs, the thermal unit *trips* to open the starter contacts, disconnecting the motor from the line. The contacts cannot be reclosed until the overload relay has been reset by moving the handle to the full off position, after allowing about 2 min for the thermal unit to cool. The open-type starter will fit into a standard outlet box and can be used with a standard flush plate. The compact construction of this type of device has the advantage of the capability of being mounted directly on the driven machinery and in various other places where the available space is small.

Manual motor-starting switches provide on-off control of single-phase or three-phase ac motors where overload protection is not required or is separately provided. Two- or three-pole switches are available with ratings up to 10 hp, 600 V, three phase. The continuous current rating is 30 A at 250 V maximum and 20 A at 600 V maximum. The toggle operation of the manual switch is similar to the fractional horsepower starter, and typical applications of the switch include small machine tools, pumps, fans, conveyors, and other electrical machinery which has separate motor protection. They are particularly suited to switching nonmotor loads, such as resistance heaters.

The integral horsepower manual starter is available in two- and three-pole versions to control single-phase motors up to 5 hp and polyphase motors up to 10 hp, respectively.

The two-pole starters have one overload relay, and three-pole starters usually have three overload relays. When an overload relay trips, the starter mechanism unlatches, opening the contacts to stop the motor. The contacts cannot be reclosed until the starter mechanism has been reset by pressing the stop button or moving the handle to the *reset* position, after allowing time for the thermal unit to cool.

Integral horsepower manual starters with low-voltage protection prevent automatic start-up of motors after a power loss. This is accomplished with a continuous-duty solenoid which is energized whenever the line-side voltage is present. If the line voltage is lost or disconnected, the solenoid deenergizes, opening the starter contacts. The contacts will not automatically close when the voltage is restored to the line. To close the contacts, the device must be manually reset. This manual starter will not function unless the line terminals are energized. This is a safety feature that can protect personnel or equipment from damage and is used on such equipment as conveyors, grinders, metalworking machinery, mixers, woodworking machinery, and wherever standards require low-voltage protection.

MAGNETIC MOTOR CONTROLLERS

Magnetic motor controllers use electromagnetic energy for closing switches. The electromagnet consists of a coil or wire placed on an iron core. When current flows through the coil, the iron of the magnet becomes magnetized, attracting the iron bar called the armature. An interruption of the current flow through the coil of wire causes the armature to drop out due to the presence of an air gap in the magnetic circuit.

Line-voltage magnetic motor starters are electromechanical devices that provide a safe, convenient, and economic means for starting and stopping motors and have the advantage of being controlled remotely. The great bulk of motor controllers are of this type. Therefore, the operating principles and applications of magnet motor controllers should be fully understood.

In the construction of a magnetic controller, the armature is mechanically connected to a set of contacts, so that when the armature moves to its closed position, the contacts also close. When the coil has been energized and the armature has moved to the closed position, the controller is said to be *picked up* and the armature *seated* or *sealed in.* Some of the magnet and armature assemblies in current use are as follows:

1. *Clapper type:* In this type, the armature is hinged. As it pivots to seal in, the movable contacts close against the stationary contacts.

2. *Vertical action:* The action is a straight-line motion with the

armature and contacts being guided so that they move in a vertical plane.

3. *Horizontal action:* Both armature and contacts move in a straight line through a horizontal plane.

4. *Bell crank:* A bell crank lever transforms the vertical action of the armature into a horizontal contact motion. The shock of armature pickup is not transmitted to the contacts, resulting in minimum contact bounce and longer contact life.

The magnetic circuit of a controller consists of the magnet assembly, the coil, and the armature. It is so named from a comparison with an electrical circuit. The coil and the current flowing in it cause magnetic flux to be set up through the iron in a manner similar to a voltage causing current to flow through a system of conductors. The changing magnetic flux produced by alternating currents results in a temperature rise in the magnetic circuit. The heating effect is reduced by laminating the magnet assembly and armature. By placing a coil of many turns of wire around a soft iron core, the magnet flux set up by the energized coil tends to be concentrated; therefore, the magnetic field effect is strengthened. Since the iron core is the path of least resistance to the flow of the magnetic lines of force, magnetic attraction will concentrate according to the shape of the magnet.

The magnetic assembly is the stationary part of the magnetic circuit. The coil is supported by and surrounds part of the magnet assembly in order to induce magnetic flux into the magnetic circuit.

The armature is the moving part of the magnetic circuit. When it has been attracted into its sealed-in position, it completes the magnetic circuit. To provide maximum pull and to help ensure quietness, the faces of the armature and the magnet assembly are ground to a very close tolerance.

When a controller's armature has sealed in, it is held closely against the magnet assembly. However, a small gap is always deliberately left in the iron circuit. When the coil becomes deenergized, some magnetic flux (residual magnetism) always remains – and if it were not for the gap in the iron circuit, the residual magnetism might be sufficient to hold the armature in the sealed-in position.

The shaded pole principle is used to provide a time delay in the decay of flux in dc coils, but it is used more frequently to prevent a chatter and wear in the moving parts of ac magnets. A shading coil is a single turn of conducting material mounted in the face of the magnet assembly or armature. The alternating main magnetic flux induces currents in the shading coil, and these currents set up auxiliary magnetic flux which is out of phase from the main flux. The auxiliary flux produces a magnetic pull out of phase from the pull due to the main flux, and this keeps the armature sealed in when the main flux fails to zero (which occurs 120 times/s with 60-cycle ac). Without the shading coil, the armature would tend to open each time the main flux goes through zero. Excessive noise, wear on magnet faces, and heat would result.

Figure 6-1 shows an exaggerated view of a pole face with a copper band or short-circuited coil of low resistance connected around a portion of the pole tip. When the flux is increasing in the pole from left to right, the induced current in the coil is in a clockwise direction.

The magnetomotive force produced by the coil opposes the direction of the flux of the main field. Therefore, the flux density in the shaded portion of the iron will be considerably less, and the flux density in the unshaded portion of the iron will be more than would be the case without the shading coil.

Figure 6-2 shows the pole with the flux still moving from left to right but decreasing in value. Now the current in the coil is in a counterclockwise direction. The magnetomotive force produced by the coil is in the same direction as the main unshaded portion. Therefore, the flux density in the shaded portion of the iron will be considerably less than it would be without the shading coil. Consequently, if the electric circuit of a coil is opened, the current decreases rapidly to zero, but the flux decreases much more slowly due to the action of the shading coil.

Electrical ratings for ac magnetic contactors and starters are shown in Fig. 6-3.

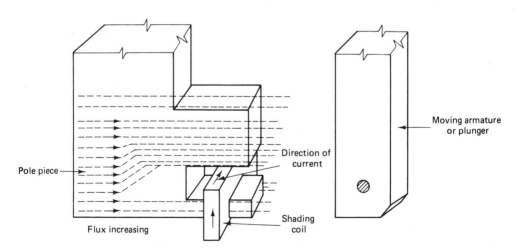

FIG. 6-1   Section of pole face with current in clockwise direction.

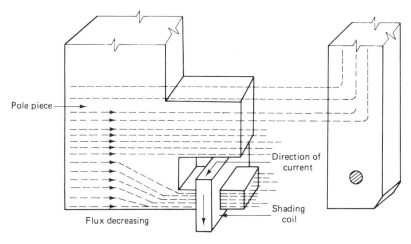

Pole piece

Direction of current

Shading coil

Flux decreasing

FIG. 6-2 Section of pole face with current in counterclockwise direction.

OVERLOAD PROTECTION

Overload protection for an electric motor is necessary to prevent burnout and to ensure maximum operating life. Electric motors will, if permitted, operate at an output of more than rated capacity. Conditions of motor overload may be caused by an overload on driven machinery, by a low line voltage, or by an open line in a polyphase system, which results in single-phase operation. Under any condition of overload, a motor draws excessive current that causes overheating. Since motor winding insulation deteriorates when subjected to overheating, there are established limits on motor operating temperatures. To protect a motor from overheating, overload relays are employed on a motor control to limit the amount of current drawn. This is overload protection, or running protection.

The ideal overload protection for a motor is an element with current-sensing properties very similar to the heating curve of the motor (see Fig. 6-4) which would act to open the motor circuit when full-load current is exceeded. The operation of the protective device should be such that the motor is allowed to carry harmless overloads but is quickly removed from the line when an overload has persisted too long.

Fuses are not designed to provide overload protection. Their basic function is to protect against short circuits (overcurrent protection). Motors draw a high inrush current when starting, and conventional single-element fuses have no way of distinguishing between this temporary and harmless inrush current and a damaging overload. Such fuses,

FIG. 6-3 Electrical ratings of ac magnetic contactors and starter. (Courtesy Sq. [D].)

NEMA Size	Load Volts	Max HP Nonplugging & Nonjogging — Single Phase	Max HP Nonplugging & Nonjogging — Poly-Phase	Max HP Plugging & Jogging — Single Phase	Max HP Plugging & Jogging — Poly-Phase	Continuous Current Rating, Amperes 600 V Max.	Service Limit Current Rating, Amperes *	Tungsten & Infrared Lamp Load, Amperes 250 V Max. ★	Resistance Heating Loads, KW ‡ — Single Phase	Resistance Heating Loads, KW ‡ — Poly-Phase	Transformers ≤20× Peak — Single Phase	Transformers ≤20× Peak — Poly-Phase	Transformers Over 20–40× Peak — Single Phase	Transformers Over 20–40× Peak — Poly-Phase	3 Phase Rating for Switching Capacitors • KVAR
00	115	1/3				9	11	5							
	200		1½			9	11	5							
	230	1	1½			9	11	5							
	380		1½			9	11								
	460		2			9	11								
	575		2			9	11								
0	115	1		½		18	21	10			0.6		0.3		
	200		3		1½	18	21	10							
	230	2	3	1	1½	18	21	10			1.2		0.6		
	380		5		1½	18	21								
	460		5		2	18	21					1.8		0.9	
	575		5		2	18	21					2.1		1.0	
1	115	2		1		27	32	15	3	5	1.2		0.6		
	200		7½		3	27	32	15		9.1		3.6		1.8	
	230	3	7½	2	3	27	32	15	6	10	2.4		1.2		
	380		10		5	27	32			16.5		4.3		2.1	
	460		10		5	27	32			20		8.5		4.3	
	575		10		5	27	32			25		11.0		5.3	
1P	115	3		1½		36	42	24							
	230	5		3		36	42	24							
2	115	3		2		45	52	30	5	8.5	2.1		1.0		
	200		10		7½	45	52	30		15.4		6.3		3.1	
	230	7½	15	5	10	45	52	30	10	17	4.1		2.1		8
	380		15		15	45	52			28		7.2		3.6	
	460		25		15	45	52			34		14		7.2	16
	575		25		15	45	52			43		18		8.9	20
3	115	7½				90	104	60	10	17	4.1		2.0		
	200		25		15	90	104	60		31		12		6.1	
	230	15	30		20	90	104	60	20	34	8.1		4.1		27
	380		30		30	90	104			56		14		7.0	
	460		50		30	90	104			68		28		14	53
	575		50		30	90	104			86		35		18	67

46

4	200	...	40	...	25	135	156	120	...	45	...	20	...	10	...
	230	...	50	...	30	135	156	120	30	52	14	23	6.8	12	40
	380	...	75	...	50	135	156	...	...	86.7	...	...	...	...	...
	460	...	100	...	60	135	156	...	60	105	27	47	14	23	80
	575	...	100	...	60	135	156	...	75	130	34	59	17	29	100
5	200	...	75	...	60	270	311	240	...	91	...	41	...	20	...
	230	...	100	...	75	270	311	240	60	105	27	47	14	24	80
	380	...	150	...	125	270	311	...	...	173	...	...	...	...	...
	460	...	200	...	150	270	311	...	120	210	54	94	27	47	160
	575	...	200	...	150	270	311	...	150	260	68	117	34	59	200
6	200	...	150	...	125	540	621	480	...	182	...	81	...	41	...
	230	...	200	...	150	540	621	480	120	210	54	94	27	47	160
	380	...	300	...	250	540	621	...	...	342	...	...	...	...	...
	460	...	400	...	300	540	621	...	240	415	108	188	54	94	320
	575	...	400	...	300	540	621	...	300	515	135	234	68	117	400
7	230	...	300	...	...	810	932	...	180	315	...	...	...	...	240
	460	...	600	...	...	810	932	...	360	625	...	...	...	...	480
	575	...	600	...	...	810	932	...	450	775	...	...	...	...	600
8	230	...	450	...	...	1215	1400	...	...	...	...	...	...	...	360
	460	...	900	...	...	1215	1400	...	...	...	...	...	...	...	720
	575	...	900	...	...	1215	1400	...	...	...	...	...	...	...	900

Tables and footnotes are taken from NEMA Standards.

† Ratings shown are for applications requiring repeated interruptions of stalled motor current or repeated closing of high transient currents encountered in rapid motor reversal, involving more than five openings or closings per minute and more than ten in a ten-minute period such as plug-stop, plug-reverse or jogging duty. Ratings apply to single speed and multi-speed controllers.

* Per NEMA Standards paragraph IC 1-21A.20, the service-limit current represents the maximum rms current, in amperes, which the controller may be expected to carry for protracted periods in normal service. At service-limit current ratings, temperature rises may exceed those obtained by testing the controller at its continuous current rating. The ultimate trip current of over-current (overload) relays or other motor protective devices shall not exceed the service-limit current ratings of the controller.

★ FLUORESCENT LAMP LOADS — 300 VOLTS AND LESS — The characteristics of fluorescent lamps are such that it is not necessary to derate Class 8502 contactors below their normal continuous current rating. Class 8903 contactors may also be used with fluorescent lamp loads. For controlling tungsten and infrared lamp loads, and resistance heating loads, Class 8903 ac lighting contactors are recommended. These contactors are specifically designed for such loads and are applied at their full rating as listed in the Class 8903 Section.

‡ Ratings apply to contactors which are employed to switch the load at the utilization voltage of the heat producing element with a duty which requires continuous operation of not more than five openings per minute. Class 8903 Types L and S lighting contactors are rated for resistance heating loads.

● When discharged, a capacitor has essentially zero impedance. For repetitive switching by contactor, sufficient impedance should be connected in series to limit inrush current to not more than 6 times the contactor rated continuous current. In many installations, the impedance of connecting conductors may be sufficient for this purpose. When switching to connect additional banks, the banks already on the line may be charged and can supply additional available short-circuit current which should be considered when selecting impedance to limit the current.

The ratings for capacitor switching above assume the following maximum available fault currents:

NEMA Size 2-3: 5,000 A RMS Sym.
NEMA Size 4-5: 10,000 A RMS Sym.
NEMA Size 6-8: 22,000 A RMS Sym.

If available fault current is greater than these values, connect sufficient impedance in series as noted in the previous paragraph.

The motor ratings in the above table are NEMA standard ratings and apply only when the code letter of the motor is the same as or occurs earlier in the alphabet than is shown in the table below.

Motors having code letters occurring later in the alphabet may require a larger controller. Consult local Square D field office.

Motor HP Rating	Maximum Allowable Motor Code Letter
1½-2	L
3-5	K
7½ & above	H

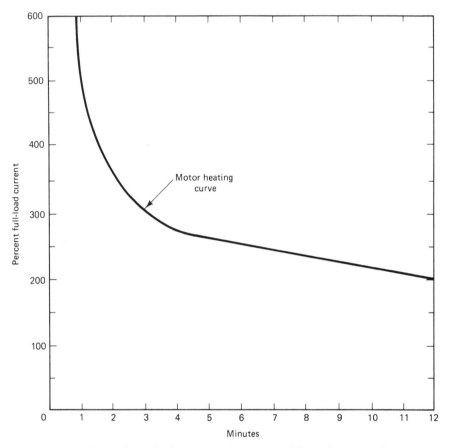

FIG. 6-4   The ideal overload protection is one in which sensing properties are very similar to the heating curve of the motor.

chosen on the basis of motor full-load current, would "blow" every time the motor is started. On the other hand, if a fuse were chosen large enough to pass the starting or inrush current, it would not protect the motor against small, harmful overloads which might occur later.

Dual element or time delay fuses can provide motor overload protection but suffer the disadvantage of being nonrenewable and must be replaced.

The overload relay is the heart of motor protection. It has inverse trip time characteristics, permitting it to hold in during the accelerating period (when inrush current is drawn) yet providing protection on small overloads above the full-load current when the motor is running. Unlike dual element fuses, overload relays are renewable and can withstand repeated trip and reset cycles without the need of replacement. They cannot, however, take the place of overcurrent protective equipment.

The overload relay consists of a current-sensing unit connected in the line to the motor, plus a mechanism, actuated by the sensing unit, which serves to directly or indirectly break the circuit. In a manual starter, an overload trips a mechanical latch, causing the starter contacts to open and disconnect the motor from the line. In magnetic starters, an overload opens a set of contacts within the overload relay itself. These contacts are wired in series with the starter coil in the control circuit of the magnetic starter. Breaking the coil circuit causes the starter contacts to open, disconnecting the motor from the line.

Overload relays can be classified as being either thermal or magnetic. Magnetic overload relays react only to current excesses and are not affected by temperature. As the name implies, thermal overload relays rely on the rising temperatures caused by the overload current to trip the overload mechanism. Thermal overload relays can be further subdivided into two types: melting alloy and bimetallic.

The melting alloy assembly of heater element (overload relay) and solder pot is shown in Fig. 6-5. Excessive overload motor current passes through the heater element, thereby melting an eutectic alloy solder pot. The ratchet wheel will then be allowed to turn in the molten pool, and a tripping action of the starter control circuit results, stopping the motor. A cooling off period is required to allow the solder pot to *freeze* before the overload relay assembly may be reset and motor service restored.

Melting alloy thermal units are interchangeable and of a one-piece construction, which ensures a constant relationship between the heater element and solder pot and allows factory calibration, making them virtually tamperproof in the field. These important features are not possible

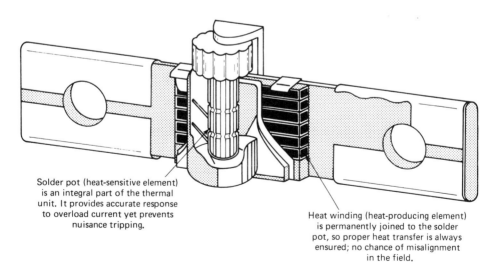

Solder pot (heat-sensitive element) is an integral part of the thermal unit. It provides accurate response to overload current yet prevents nuisance tripping.

Heat winding (heat-producing element) is permanently joined to the solder pot, so proper heat transfer is always ensured; no chance of misalignment in the field.

FIG. 6-5   Melting alloy thermal overload relay.

with any other type of overload relay construction. A wide selection of these interchangeable thermal units is available to give exact overload protection of any full-load current to a motor.

Bimetallic overload relays are designed specifically for two general types of application: The automatic reset feature is of decided advantage when devices are mounted in locations not easily accessible for manual operation, and, second, these relays can easily be adjusted to trip within a range of 85–115% of the nominal trip rating of the heater unit. This feature is useful when the recommended heater size might result in unnecessary tripping, while the next larger size would not give adequate protection. Ambient temperatures affect overload relays operating on the principle of heat.

Ambient-compensated bimetallic overload relays were designed for one particular situation, that is, when the motor is at a constant temperature and the controller is located separately in a varying temperature. In this case, if a standard thermal overload relay were used, it would not trip consistently at the same level of motor current if the controller temperature changed. This thermal overload relay is always affected by the surrounding temperature. To compensate for the temperature variations the controller may see, an ambient-compensated overload relay is applied. Its trip point is not affected by temperature, and it performs consistently at the same value of current.

Melting alloy and bimetallic overload relays are designed to approximate the heat actually generated in the motor. As the motor temperature increases, so does the temperature of the thermal unit. The motor and relay heating curves (see graph in Fig. 6-6) show this relationship. From this graph we can see that no matter how high the current drawn, the overload relay will provide protection, yet the relay will not trip out unnecessarily. See Figs. 6-7 and 6-8.

When selecting thermal overload relays, the following must be considered:

1.   Motor full-load current

2.   Type of motor

3.   Difference in ambient temperature between motor and controller

Motors of the same hp and speed do not all have the same full-load current, and the motor nameplate must always be referred to to obtain the full-load amps for a particular motor. Do not use a published table. Thermal unit selection tables are published on the basis of continuous-duty motors, with a 1.15 service factor, operating under normal conditions. The tables are shown in the catalog of manufacturers and also appear on the inside of the door or cover of the motor controller. These selections

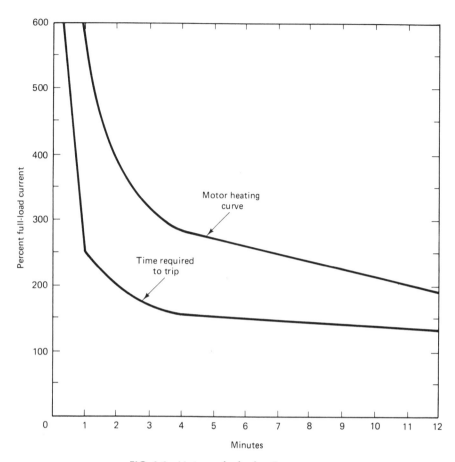

FIG. 6-6   Motor and relay heating curves.

will properly protect the motor and allow the motor to develop its full hp, allowing for the service factor, if the ambient temperature is the same at the motor as at the controller. If the temperatures are not the same or if the motor service factor is less than 1.15, a special procedure is required to select the proper thermal unit.

Standard overload relay contacts are closed under normal conditions and open when the relay trips. An alarm signal is sometimes required to indicate when a motor has stopped due to an overload trip. Also, with some machines–particularly those associated with continuous processing–it may be required to signal an overload condition rather than have the motor and process stop automatically. This is done by fitting the overload relay with a set of contacts which close when the relay trips, so completing the alarm circuit. These contacts are appropriately called *alarm contacts*.

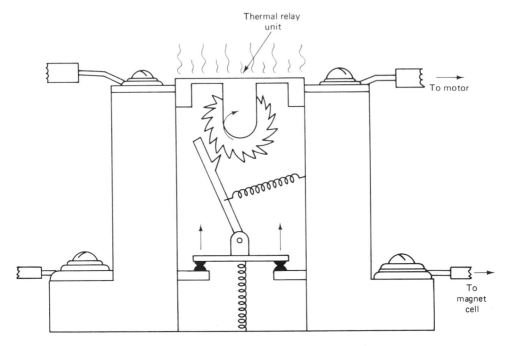

FIG. 6-7 Operation of melted alloy overload relay.

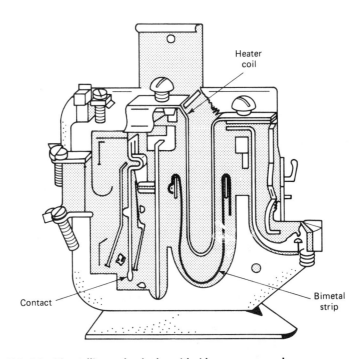

FIG. 6-8 Bimetallic overload relay with side cover removed.

A magnetic overload relay has a movable magnetic core inside a coil which carries the motor current. The flux set up inside the coil pulls the core upward. When the core rises far enough, it trips a set of contacts on the top of the relay. The movement of the core is slowed by a piston working in an oil-filled dash pot mounted below the coil. This produces an inverse-time characteristic. The effective tripping current is adjusted by moving the core on a threaded rod. The tripping time is varied by uncovering oil bypass holes in the piston. Because of the time and current adjustments, the magnetic overload relay is sometimes used to protect motors having long accelerating times or unusual duty cycles.

PROTECTIVE ENCLOSURES

The correct selection and installation of an enclosure for a particular application can contribute considerably to the length of life and trouble-free operation. To shield electrically live parts from accidental contact, some form of enclosure is always necessary. This function is usually fulfilled by a general-purpose, sheet steel cabinet. Frequently, however, dust, moisture, or explosive gases make it necessary to employ a special enclosure to protect the motor controller from corrosion or the surrounding equipment from explosion. In selecting and installing control apparatus, it is always necessary to consider carefully the conditions under which the apparatus must operate; there are many applications where a general-purpose enclosure does not afford protection.

The Underwriters' Laboratories have defined the requirements for protective enclosures according to the hazardous conditions, and the National Electrical Manufacturers Association has standardized enclosures from these requirements:

NEMA 1 – General purpose: The general-purpose enclosure is intended primarily to prevent accidental contacts with the enclosed apparatus. It is suitable for general-purpose applications indoors where it is not exposed to unusual service conditions. A NEMA 1 enclosure serves as protection against dust and light indirect splashing but is not dust-tight.

NEMA 3 – Dusttight, raintight: This enclosure is intended to provide suitable protection against specified weather hazards. A NEMA 3 enclosure is suitable for application outdoors, on ship docks, on canal locks, on construction work, and in subways and tunnels. It is also sleet resistant.

NEMA 3R – Rainproof, sleet resistant: This enclosure protects against interference in operation of the contained equipment due to rain and resists damage from exposure to sleet. It is designed with conduit hubs and external mounting as well as drainage provisions.

NEMA 4 – Watertight: A watertight enclosure is designed to meet a hose test which consists of a stream of water from a hose with a 1-in. nozzle, delivering at least 65 gal/min. The water is directed on the enclosure from a distance of not less than 10 ft and for a period of 5 min. During this period, it may be directed in any one or more directions as desired. There shall be no leakage of water into the enclosure under these conditions. Such an enclosure is suitable for applications outdoors on ship docks and in dairies, breweries, and the like.

NEMA 4X – Watertight, corrosion resistant: These enclosures are generally constructed along the lines of NEMA 4 enclosures except they are made of a material that is highly resistant to corrosion. For this reason, they are ideal in applications such as paper mills and meat packing, fertilizer, and chemical plants where contaminants would ordinarily destroy a steel enclosure over a period of time.

NEMA 7 – Hazardous locations – class I: These enclosures are designed to meet the application requirements of the National Electrical Code for class I hazardous locations. In this type of equipment, the circuit interruption occurs in air.

Class I locations are those in which flammable gases or vapors are or may be present in the air in quantities sufficient to produce explosive or ignitible mixtures.

NEMA 9 – Hazardous locations – class II: These enclosures are designed to meet the application requirements of the NE Code for class II hazardous locations.

Class II locations are those which are hazardous because of the presence of combustible dust.

The letter or letters following the type number indicates the particular group or groups of hazardous locations (as defined in the NE Code) for which the enclosure is designed. The designation is incomplete without a suffix letter or letters.

NEMA 12 – Industrial use: This type of enclosure is designed for use in those industries where it is desired to exclude such materials as dust, lint, fibers and flyings, oil seepage, or coolant seepage. There are no conduit openings or knockouts in the enclosure, and mounting is by means of flanges or mounting feet.

NEMA 13 – Oiltight, dusttight: NEMA 13 enclosures are generally made of cast iron, gasketed to permit use in the same environments as NEMA 12 devices. The essential difference is that, due to its cast housing a conduit entry is provided as an integral part of the NEMA 13 enclosure, and mounting is by means of blind holes rather than mounting brackets.

NATIONAL ELECTRICAL CODE REQUIREMENTS

The National Electrical Code deals with the installation of equipment and is primarily concerned with safety – the prevention of injury and fire hazard to persons and property arising from the use of electricity. It is adopted on a local basis, sometimes incorporating minor changes or interpretations, as the need arises. NE Code rules and provisions are enforced by governmental bodies exercising legal jurisdiction over electrical installations and used by insurance inspectors. Minimum safety standards are thus assured.

Motor-control equipment is designed to meet the provisions of the NE Code. Code sections applying to the industrial control devices are Article 430 on motors and motor controllers and Article 500 on hazardous locations.

With minor exceptions, the NE Code, along with some local codes, requires a disconnect means for every motor. A combination starter consists of an across-the-line starter and a disconnect means wired together in a common enclosure. Combination starters include a blade-disconnect switch, either fusible or nonfusible, while some combination starters include a thermal-magnetic trip circuit breaker. The starter may be controlled remotely with push buttons, selector switches, and the like, or these devices may be installed in the cover. The single device makes a neat as well as compact electrical installation that takes little mounting space.

A combination starter provides safety for the operator, because the cover of the enclosing case is interlocked with the external operating handle of the disconnecting means. The door cannot be opened with the disconnecting means closed. With the disconnect means open, there can be access to all parts, but much less hazard is involved inasmuch as there are no readily accessible parts connected to the power line. This safety feature cannot be obtained with separately enclosed starters. In addition, the cabinet is provided with a means for padlocking the disconnect in the off position.

CONTROL CIRCUITS

TWO-WIRE CONTROL.   Figure 6-9 shows wiring diagrams for a two-wire control circuit. The control itself could be a thermostat, float switch, limit switch, or other maintained contact device to the magnetic starter. When the contacts of the control device close, they complete the coil circuit of the starter, causing it to pick up and connect the motor to the lines. When the control device contacts open, the starter is deenergized, stopping the motor.

Two-wire control provides low-voltage release but not low-voltage

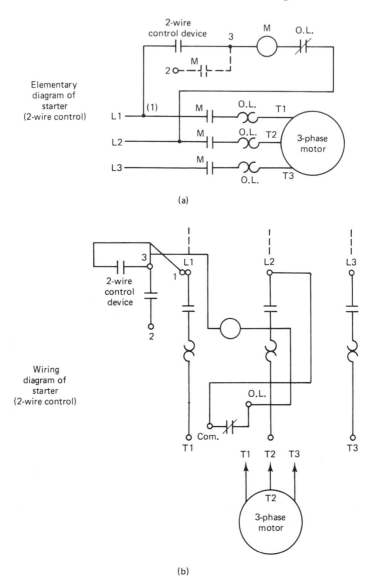

FIG. 6-9   Elementary diagram of two-wire motor control.

protection. When wired as illustrated, the starter will function automatically in response to the direction of the control device, without the attention of an operator. In this type of connection, a holding circuit interlock is not necessary.

THREE-WIRE CONTROL. A three-wire control circuit uses momentary contact, start-stop buttons, and a holding circuit interlock wired in parallel with the start button to maintain the circuit. Pressing the normally open (N.O.) start button completes the circuit to the coil. The power circuit contacts in lines 1, 2, and 3 close, completing the circuit to the motor, and the holding circuit contact also closes. Once the starter has picked up, the start button can be released, as the now-closed interlock contact provides an alternate current path around the reopened start contact.

Pressing the normally closed (N.C.) stop button will open the circuit to the coil, causing the starter to drop out. An overload condition, which causes the overload contact to open, a power failure, or a drop in voltage to less than the seal-in value, would also deenergize the starter. When the starter drops out, the interlock contact reopens, and both current paths to the coil, through the start button and the interlock, are now open.

Since three wires from the push-button station are connected into the starter at points 1, 2, and 3, this wiring scheme is commonly referred to as three-wire control. See Fig. 6-10.

The holding circuit interlock is a normally open (N.O.) auxiliary contact provided on standard magnetic starters and contactors. It closes when the coil is energized to form a holding circuit for the starter after the *start* button has been released.

In addition to the main or power contacts which carry the motor current and the holding circuit interlock, a starter can be provided with externally attached auxiliary contacts, commonly called electrical interlocks. Interlocks are rated to carry only control circuit currents, not motor currents. Both N.O. and N.C. versions are available. Among a wide variety of applications, interlocks can be used to control other magnetic devices where sequence operation is desired, to electrically prevent another controller from being energized at the same time, and to make and break circuits to indicating or alarm devices such as pilot lights, bells, or other signals.

The circuit in Fig. 6-11 shows a three-pole reversing starter used in controlling a three-phase motor. Three-phase squirrel cage motors can be reversed by reconnecting any two of the three line connections to the motor. By interwiring two contactors, an electromagnetic method of making the reconnection can be obtained.

As shown in the power circuit (Fig. 6-12), the contacts of the forward contactor – when closed – connect lines 1, 2, and 3 to the motor terminals $T1$, $T2$, and $T3$, respectively. As long as the forward contacts are closed, mechanical and electrical interlocks prevent the reverse contactor from being energized.

When the forward contactor is deenergized, the second contactor can be picked up, closing its contacts which reconnect the lines to the motor. Note that by running through the reverse contacts, line 1 is connected to motor terminal $T3$, and line 3 is connected to motor terminal $T1$. The motor will now run in reverse.

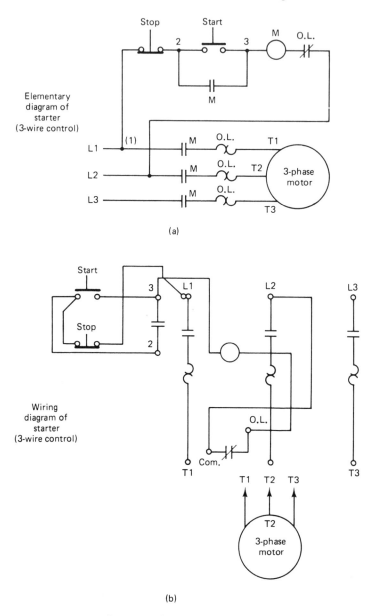

FIG. 6-10   Elementary diagram of three-wire motor control.

Manual reversing starters (employing two manual starters) are also available. As in the magnetic version, the forward and reverse switching mechanisms are mechanically interlocked, but since coils are not used in the manually operated equipment, electrical interlocks are not furnished.

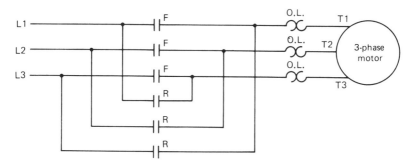

FIG. 6-11   Diagram of a three-pole reversing starter used to control a three-phase motor.

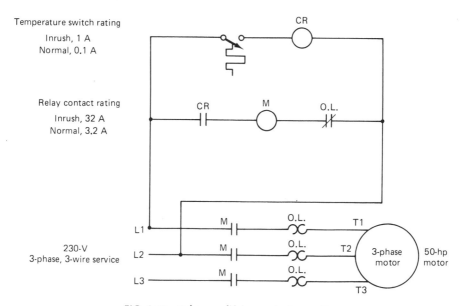

FIG. 6-12   Relay amplifying contact capacity.

CONTROL RELAYS.   A relay is an electromagnetic device whose contacts are used in control circuits of magnetic starters, contactors, solenoids, timers, and other relays. They are generally used to amplify the contact capability or multiply the switching functions of a pilot device.

The wiring diagrams in Fig. 6-13 demonstrate how a relay amplifies contact capacity. Figure 6-13(a) represents a current amplification. Relay and starter coil voltages are the same, but the ampere rating of the temperature switch is too low to handle the current drawn by the starter coil. A relay is interposed between the temperature switch and starter coil.

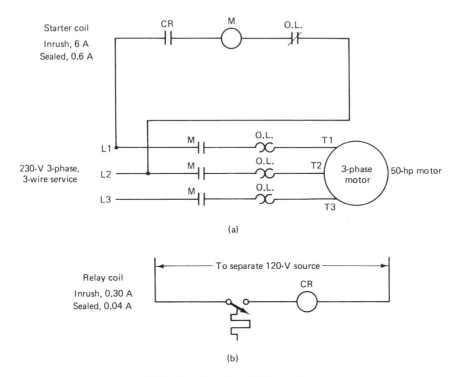

Starter coil
Inrush, 6 A
Sealed, 0.6 A

230-V 3-phase,
3-wire service

(a)

Relay coil
Inrush, 0.30 A
Sealed, 0.04 A

(b)

FIG 6-13   Circuit amplifying voltage.

The current drawn by the relay coil (CR) is within the rating of the temperature switch, and the relay contact (CR) has a rating adequate for the current drawn by the starter coil.

The other drawing [Fig. 6-13(b)] represents a voltage amplification. A condition may exist in which the voltage rating of the temperature switch is too low to permit its direct use in a starter control circuit operating at some higher voltage. In this application, the coil of the interposing relay and the pilot device are wired to a low-voltage source of power compatible with the rating of the pilot device. The relay contact, with its higher voltage rating, is then used to control the operation of the starter.

Relays are commonly used in complex controllers to provide the logic or "brains" to set up and initiate the proper sequencing and control of a number of interrelated operations.

In selecting a relay for a particular application, one of the first steps should be a determination of the control voltage at which the relay will operate. Once the voltage is known, the relays which have the necessary contact rating can be further reviewed and a selection made on the basis of the number of contacts and other characteristics needed.

OTHER CONTROLLING EQUIPMENT

TIMERS AND TIMING RELAYS. A pneumatic timer or timing relay is similar to a control relay except that certain of its contacts are designed to operate at a preset time interval after the coil is energized or deenergized. A delay on energization is also referred to as *on delay*. A time delay on deenergization is also called *off delay*.

A timed function is useful in applications such as the lubricating system of a large machine in which a small oil pump must deliver lubricant to the bearings of the main motor for a set period of time before the main motor starts.

In pneumatic timers, the timing is accomplished by the transfer of air through a restricted orifice. The amount of restriction is controlled by an adjustable needle valve, permitting changes to be made in the timing period.

DRUM SWITCH. A drum switch is a manually operated three-position, three-pole switch which carries an hp rating and is used for manual reversing of single- or three-phase motors. Drum switches are available in several sizes and can be spring-return-to-off (momentary contact) or maintained contact. Separate overload protection, by manual or magnetic starters, must usually be provided, as drum switches do not include this feature.

PUSH-BUTTON STATION. A control station may contain push buttons, selector switches, and pilot lights. Push buttons may be momentary or maintained contact. Selector switches are usually maintained contact but they can be spring-return to give momentary contact operation.

Standard-duty stations will handle the coil currents of contactors up to size 4. Heavy-duty stations have higher contact ratings and provide greater flexibility through a wider variety of operators and interchangability of units.

FOOT SWITCH. A foot switch is a control device operated by a foot pedal used where the process or machine requires that the operator have both hands free. Foot switches usually have momentary contacts but are available with latches which enable them to be used as maintained contact devices.

LIMIT SWITCH. A limit switch is a control device which converts mechanical motion into an electrical control signal. Its main function is to limit movement, usually by opening a control circuit when the limit of travel is reached. Limit switches may be momentary contact (spring return) or maintained contact types. Among other applications, limit switches can be used to start, stop, reverse, slow down, speed up, or recycle machine operations.

SNAP SWITCH. Snap switches for motor-control purposes are enclosed, precision switches which require low operating forces and have a high repeat accuracy. They are used as interlocks and as the switch mechanism for control devices such as precision limit switches and pressure switches. They are available also with integral operators for use as compact limit switches, door-operated interlocks, etc. Single-pole, double-throw, and two-pole double-throw versions are available.

PRESSURE SWITCH. The control of pumps, air compressors, welding machines, lube systems, and machine tools requires control devices which respond to the pressure of a medium such as water, air, or oil. The control device which does this is a pressure switch. It has a set of contacts which are operated by the movement of a piston, bellows, or diaphragm against a set of springs. The spring pressure determines the pressures at which the switch closes and opens its contacts.

FLOAT SWITCH. When a pump motor must be started and stopped according to changes in water (or other liquid) level in a tank or sump, a float switch is used. This is a control device whose contacts are controlled by movement of a rod or chain and counterweight, fitted with a float. For closed tank applications, the movement of a float arm is transmitted through a bellows seal to the contact mechanism.

ELECTRONIC CONTROLS

In recent years, many solid-state devices and circuits have entered the motor control field, reducing previously bulky equipment to compact, efficient, and reliable electronic units. Even electric motors have been built on printed circuit boards with all windings made from flimsy copper foil mounted on a flat card.

The Square D NORPAK® Solid State Logic Control is one type of control system that has recently been introduced. The NOR is the basic logic element for NORPAK. Using this single element allows many functions to be obtained through the building block approach. While this results in simplicity of design and the minimum of logic elements in a system, it is also a great benefit when changes are required. While NORs can be connected to form any logic function, it is sometimes more convenient to have other logic functions available. For this reason, ANDs, sealed ANDs, ORs, and memories are available. Other functions that cannot be made up from NORs alone are timers, single shots, transfer memories, counters, and shift registers. This variety of components makes this system a wise choice for any application from simple machine control up to complex systems requiring counting and data manipulation. The basic design of a NORPAK system is the same whether plug-in or encapsulated components are used.

ENERGY MANAGEMENT SYSTEM

The increasing cost of electricity and the shortage of fuel are major items of concern for top management today. Utility bills have risen to levels where action must be taken to help maintain a facility's profitability.

Most commercial and industrial electric bills are made up of two charges, energy and demand. The energy charge is based on the quantity of energy consumed for the billing period, while in most cases the demand charge is based on the peak electrical energy used during short periods of time called demand intervals.

One possible solution to lowering these demand and energy costs is through the use of an energy management system, which is a technique of automatically controlling the demand and energy consumption of a facility to a lower and more economical level by shedding and cycling noncritical loads for brief periods of time. This can be accomplished by using demand controllers with their basic features or, where economically feasible, by using complex computer-based systems. In some applications, these larger systems are justified, but in many cases smaller, less expensive controllers will do the job.

With the advent of the microprocessor, the features of the demand controller along with many of the added features of the computer-based system have been combined into the Class 8865 EM WATCHDOG Energy Management Systems. The controllers use a continuously integrating demand control technique. It is based on an electronic version of the conventional thermal lagged-demand meter. Loads are shed when the predicted demand equals the programmed demand limit.

The priorities, shed and restore times, along with system data can be planned on a data entry worksheet. A calculator-type keyboard is used to enter worksheet data into the controller. A digital readout display confirms that data have been entered.

If a programmed error is made, a digital readout will display an error code. To correct an error, the clear entry key is touched, and the correct data are reentered. All programming data may then be easily entered or changed.

After all data have been entered, a key switch can be locked into the RUN position which prevents unauthorized persons from changing critical load data. While in the RUN mode, the controller's digital readout will automatically display the predicted demand, the demand limit, and the time of day. All other data settings can be displayed upon request. A battery backup is provided to retain stored information in the event of a power failure.

PROGRAMMABLE CONTROLLERS

Programmable controllers may be used in place of conventional control; they include relays offering faster start-ups, decreased start-up costs, quick program changes, fast troubleshooting, and up-to-date schematic diagram printouts. They also offer additional benefits such as precision digital timing, counting, data manipulation, remote inputs and outputs, redundant control, supervisory control, process control, management report generation, machine cycle, controller self-diagnostics, dynamic graphic displays, and so forth. Applications include machine tool, sequential, process, conveyor, batching, and energy management control applications. The manufacturer should be contacted for a complete description and operation of these systems as even a simple explanation would take more space than can be allotted in this book.

7

Direct-Current
Armature Windings

Armature windings can be divided into two general classes according to the methods of connecting the coils to the commutator. They are called lap windings and wave windings, the names of which are derived from the appearance of the coils when they are traced through the winding.

LAP WINDINGS AND WAVE WINDINGS

Figure 7-1 shows a section of a lap winding. Starting with the coil at the left, trace the path of current through this coil as shown by the arrows and then on through the next coil, etc., until the end of the circuit is reached. You will find all the coils to be alike, but to clarify matters, the coil to the extreme left is drawn with heavier lines to make the tracing easier. Note that each coil overlaps the next as it is traced—thus the name *lap* windings, since the coils overlap.

The winding in Fig. 7-2 is the wave type. Again, start at the left and trace the path of current through the two coils shown by the heavy lines. Note the location of the north and south field poles, which are shown by the dotted rectangles and marked N and S. In tracing the circuit through, it can be seen that each coil in this circuit is separated from the last by the distance of one pair of poles, which gives a wave-like appearance of the two coils traced in the heavy lines; from this appearance comes the name *wave* winding.

A lap winding is a parallel winding with an even number of seg-

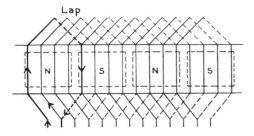

FIG. 7-1 Sectional view of a lap winding. (Courtesy Page Power Company.,

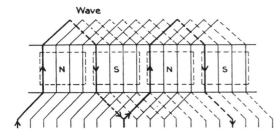

FIG. 7-2 Sectional view of a wave winding. (Courtesy Page Power Company.)

ments and slots. The number of bars, however, is equal to, or a multiple of, the number of slots – depending on the number of coils per slot. The term parallel winding comes from the fact that there are as many circuits in parallel as there are poles. Thus, a four-pole motor has four circuits in parallel and a six-pole motor has six circuits in parallel through the armature. This requires the same number of brushes as there are poles for the motor.

Since there are so many paths in parallel in a lap winding, a large amount of current can pass through the winding. This, in turn, reduces the amount of voltage needed in the armature to "push" the current through so many paths.

The lead connections for a lap winding come from each side of the coil as top and bottom leads and connect adjacent to each other on the commutator. The position that they connect on the commutator cannot be set down to a hard and fast theoretical rule but is at the discretion of the manufacturer who designs a certain motor.

There are three groups of wave winding, namely, simplex wave winding, duplex wave winding, and triplex wave winding. Wave windings are known as series windings and are generally used for motors of higher voltage and smaller currents.

Both lap and wave windings are used in armatures from fractional horsepower sizes to those of hundreds of horsepower. The type of winding selected by the motor designer depends on several factors in the electrical and mechanical requirements of the machine. Wave windings require only two brushes on the commutator, while lap windings have as many brushes as there are field poles. The wave winding permits the brushes to be located at adjacent poles and also on whichever side of the commutator they may be most convenient and accessible for inspection and repairs.

The major differences between lap and wave windings include the following:

1. A lap winding has half the voltage rating of a wave-wound armature, wound identically, and therefore can carry much more current through it.

2. The span of the leads of a lap winding, being adjacent, is not dependent on the number of poles, as is the lead span of a wave winding, where the leads are widely separated.

3. In a wave winding there are only half as many brushes as there are in a lap winding because the current has only half as many paths.

An armature winding is an electromagnet having a number of coils connected to commutator bars. There must be at least one start and one finish lead connected to each commutator bar. The coil leads of a lap-wound armature connect to commutator bars that are near each other, and the coil leads of a wave-wound armature connect to commutator bars that are widely separated. See Figs. 7-3 and 7-4.

When current flows through the coil in a clockwise direction, a south pole will be produced on the surface of the armature. If the current flows in a counterclockwise direction, a north pole will be produced on the surface of the armature. A large number of coils are used to produce a strong magnetic pole and a smoother twisting action.

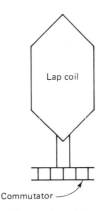

FIG. 7-3 Connection of lap winding to commutator.

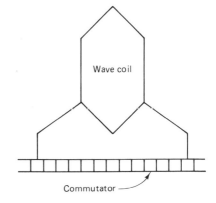

FIG. 7-4 Connection of wave winding to commutator.

Although there are only two types of dc armature windings, there are a number of winding connections that apply to either a lap- or a wave-wound armature.

SYMMETRICAL AND NONSYMMETRICAL CONNECTIONS. If the coil leads connect to commutator bars that are on a line with the center of the coil, the connection is symmetrical (Fig. 7-5). If the coil leads connect to commutator bars that are not on a line with the center of the coil, the connection is nonsymmetrical (Fig. 7-6).

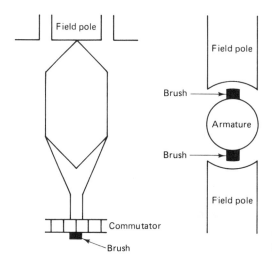

FIG. 7-5 When coil leads are connected as shown here, the connection is symmetrical.

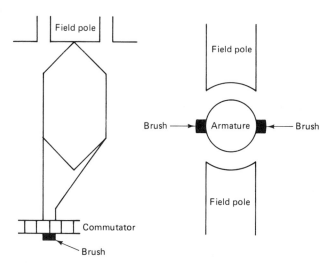

FIG. 7-6 When coil leads connect to commutator bars that are not on line with the center of the coil, the connection is nonsymmetrical.

The brushes must always short the coil when it is in the neutral plane, which means that the brushes must be located on a line with the center of the field pole if the coil is connected symmetrically and located between the field poles if connected nonsymmetrically.

PROGRESSIVE AND RETROGRESSIVE CONNECTIONS.   If the start and finish leads of a coil, or the element of a coil, do not cross the connection, it is known as progressive as shown in Fig. 7-7. If the start and finish leads of a coil, or the element of a coil, cross the winding, it is connected retrogressively. See Fig. 7-8.

If a winding is changed from progressive to retrogressive or vice versa, the effect will be reversed rotation of a motor and reversed brush polarity on a generator. Lap-wound armatures are usually connected progressively and wave-wound armatures retrogressively.

ELEMENT WINDINGS.   Element windings are used to reduce the voltage across adjacent commutator bars and decrease the tendency of brush sparking. *Example:* An armature has 30 turns/coil, and the voltage per turn is 1 or 30 V/coil. If the coil is wound in one section and connected to adjacent commutator bars, the voltage across the bars will be 30 V. Such a coil would have one start and one finish lead, and there would be as many bars as slots. This would be a single-element winding as shown in Fig. 7-9.

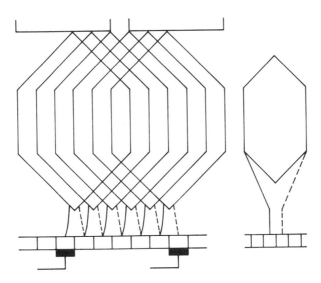

FIG. 7-7   If the start and finish leads of a coil do not cross the connection, it is known as *progressive*.

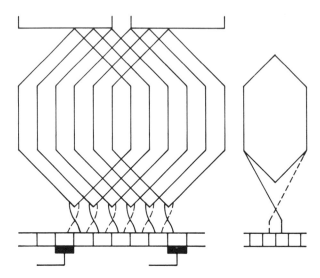

FIG. 7-8 If the start and finish leads cross the winding, it is connected
retrogressively.

If this coil were divided into sections (15 turns/section) and each section were connected to adjacent bars, the voltage across adjacent bars
would be 15 V. Such a coil would have two start and two finish leads, and
there would be twice as many bars as slots. This would be known as a
two-element winding as shown in Fig. 7-10.

Should the coil be divided in three sections (10 turns/section) and
each section connected to adjacent bars, the voltage across adjacent bars
would be 10 V. Such a coil would have three start and three finish leads,
and there would be three times as many bars as slots. This would be
known as a three-element winding. See Fig. 7-11.

Element windings are particularly desirable for high-voltage machines. The practical limit is usually three or four elements. On high-
voltage machines the voltage between bars usually does not exceed about
25 V. On smaller machines it may range form 2 to 10 V. Therefore, the
higher the voltage the machine is to be operated at, the greater number of
commutator bars it will usually have.

The number of slots in an armature is determined by the number of
poles and the practical number of slots which can be used per pole. The
slots, of course, cannot be too numerous or close together, or there will
not be sufficient iron between the coils to provide a good magnetic path
through the armature for the field flux.

The number of slots is generally considered in determining the exact
number of commutator bars, as the number of bars is usually a multiple
of the number of slots. For example, an armature with 24 slots might

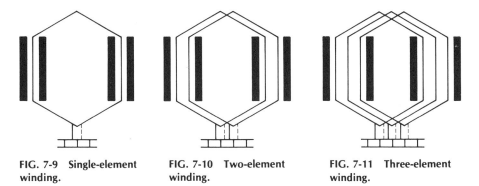

FIG. 7-9   Single-element
winding.

FIG. 7-10   Two-element
winding.

FIG. 7-11   Three-element
winding.

have 24, 48, or 72 commutator bars. In the last case the coils would be wound with three conductors in parallel, and the six leads from each coil would be connected to the proper bars.

Armature windings, therefore, can be called single-element, double-element, or three-element windings according to the number of conductors in parallel in the coils and the number of bars in proportion to the number of slots.

ARMATURE EQUALIZER CONNECTIONS

Equalizer connections provide better commutation, make possible one half of the number of brushes usually used on the lap-wound machine, and provide the manufacturer with a means of avoiding the special slot and commutator bar relationships demanded by wave-type windings. Inasmuch as the equalizers here referred to are permanently connected to the commutator and inasmuch as they make testing of the armature impossible by the regular procedure, the testing method and other information about these connections should prove of value to maintenance electricians and armature shop workers.

The principal purpose of equalizers is to connect together on the armature those points which have the same polarity and which should have equal potential. See Fig. 7-12. For a four-pole winding this means commutator bars 180° apart; for a six-pole armature, bars 120° apart; and for an eight-pole machine, bars 90° apart. The number of bars spanned by the equalizer will equal bars divided by pairs of poles. For the armature shown in Fig. 7-12 each equalizer will span 24/2 or 12 bars, thereby making the connection 1 and 13, 2 and 14, etc. The pitch for any other number of bars or poles would be determined by the same method.

To test such an armature, current must be fed to the armature from an external low-voltage dc supply, such as a battery, the leads being connected to commutator segments one half of the equalizer pitch apart.

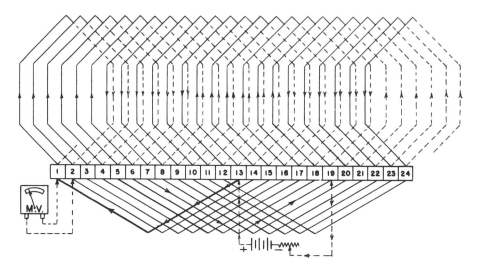

FIG. 7-12   Armature equalizer connection. (Courtesy Page Power Co.)

Since the equalizer pitch is 12 segments in this case, the leads will be spaced six bars apart or 1 and 7. Any pair of bars so spaced may be used in a fully equalized armature, bars 13 and 15 being employed in the diagram in Fig. 7-12.

The value of the test current is adjusted to give satisfactory deflection on the millivoltmeter, and the voltage drop readings are taken between all adjacent pairs of segments. These readings are interpreted in the usual manner, low readings indicating shorts and high readings showing high-resistance connections or opens. By tracing the winding and also by actual test, it will be noted that if the readings from bars 13 and 19 are forward, then the readings from 19 to 1 will be backward, 1 to 7 will be forward, and 1 to 13 backward. This is a normal indication obtained in all windings.

If the factors mentioned are kept in mind, the procedure given will produce consistently accurate results. It is to be noted such an armature will, when tested on a growler, give a shorted indication on all coils, even though the winding is in perfect condition. The reason for this can be seen by tracing from bar 1 through the coil to bar 2, through the equalizer to bar 14, through the coil to bar 13, and back through the equalizer to bar 2. Thus every coil on the armature is apparently short-circuited by having another coil placed in series with it through the equalizer connections. This explains the need for a special testing procedure.

WINDING SMALL ARMATURES

To understand how a small dc motor armature is wound, a step-by-step description should prove helpful. A small two-pole, two-element, nonsymmetrical armature having 12 slots and 24 segments will be used as an example.

The slots should first be lined with fish paper about 7 to 10 mils thick and varnished cambric about 7 mils thick. The fish paper is placed in the slot first, next to the iron core, and the varnished cloth or cambric is placed inside the fish paper. To complete the insulation of the core, use a fiber lamination at each end, which is shaped the same as the iron core laminations and has the same number of slots stamped in it. This protects the coils at the corners of the slots.

The armature should be held or clamped with the commutator end next to the winder.

In winding the first coil, the number of turns will depend on the size of the armature and its voltage rating. If this number is taken from coils in an old winding, the turns in one or more of the old coils should be very carefully counted.

When winding an armature that has twice as many bars as slots, wind two coils in each slot, thereby providing enough coil leads for all bars.

The first coils for this armature will go in slots 1 and 7, winding to the right of the shaft, at both the front and back ends of the core. After winding in one coil, a loop about 4 in. long should be made at slot 1. Then continue and wind the same number of turns again, still in slots 1–7. When the last turn is finished, run the wire from the seventh slot over to the second and make a loop at slot 2. Next wind a coil in slots 2 and 8 and again make another loop at slot 2. Then place another coil in the same slots 2 and 8 and finish with a loop at slot 3, etc. This places two coils and two loops in each slot, and the same procedure should be followed until there are two coils and two loops in every slot.

The slot insulation should then be folded over the tops of the coils and the wedges driven in.

The loops are next connected to the commutator, one loop to each segment, and they should be connected in the same way they were made in the winding. That is, the first and last single wires are brought together and connected to a segment straight out from the first slot. The second loop in the first slot is connected to the next bar, and the first loop in the second slot is connected to the next, etc.

To avoid mistakes, these loops should be marked with cotton sleeving which is slipped on over them as they are made. Red sleeving could be used on the first loop of each slot and white sleeving on the second, which will make it easy to locate the first and second loops for each slot. This winding would be used in a two-pole frame and has two circuits with 12 coils in each. If 120 V were applied to this winding, the voltage between adjacent commutator segments would be 120/12 or 10 V, which is not too high between adjacent bars. If this same armature had a commutator of only 12 segments, the voltage between bars would be 120/6 or 20 V, which is a little high for an armature of this size.

A typical lap winding for a two-pole dc motor is shown in Fig. 7-13, while various armature winding connections are shown in Fig. 7-14.

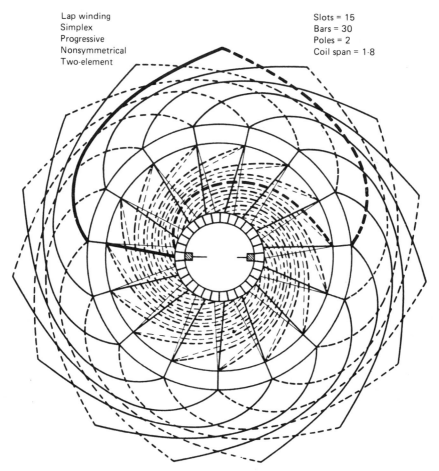

Lap winding
Simplex
Progressive
Nonsymmetrical
Two-element

Slots = 15
Bars = 30
Poles = 2
Coil span = 1-8

FIG. 7-13   Typical lap winding for a two-pole dc motor. (Courtesy Page Power Co.)

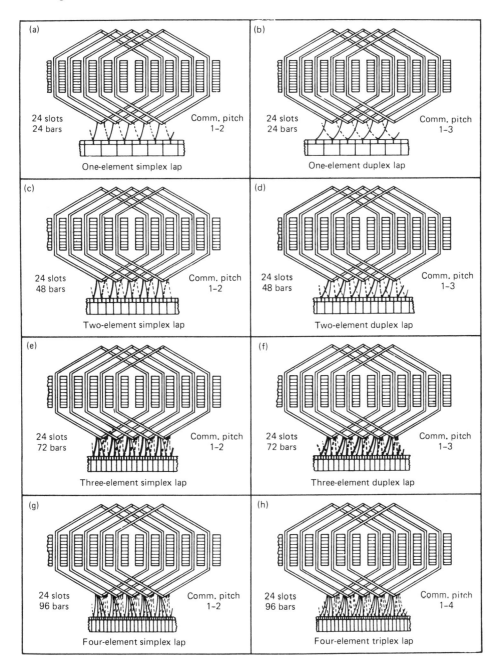

FIG. 7-14 Various armature winding connections. (Courtesy Page Power Co.)

8

Direct-Current Motors

A direct-current motor is a machine for converting dc electrical energy into rotating mechanical energy. The principle underlying operation of a dc motor is called *motor actions*, based on the fact that when a wire carrying current is placed in a magnetic field, a force is exerted on the wire, moving it through the magnetic field. There are three elements of motor action as it takes place in a dc motor:

1. Many coils of wire are wound on a cylindrical rotor or armature on the shaft of the motor.

2. A magnetic field necessary for motor action is created by placing fixed electromagnetic poles around the inside of the cylindrical motor housing. When current is passed through the fixed coils, a magnetic field is set up within the housing. Then, when the armature is placed inside the motor housing, the wires of the armature coils will be situated in the field of magnetic lines of force set up by the electromagnetic poles arranged around the stator. The stationary cylindrical of the motor is called the stator.

3. The shaft of the armature is free to rotate because it is supported at both ends by bearing brackets. Freedom of rotation is assured by providing clearance between the rotor and the faces of the magnetic poles.

DC MOTOR CONSTRUCTION

The frame (Fig. 8-1) is made of iron to be used to complete the magnetic circuit for the field poles. Traditionally, frames have been made in three types: open, semienclosed, and closed types. The open frame has the end plates or bells open so the air can freely circulate through the machine. The semienclosed frame has a wire netting or small holes in the end bells so that air can enter but will prevent any foreign material entering the machine. The enclosed-type frame has the end bells completely closed, and the machine is airtight. The closed-type frame is used in certain areas such as those classified as hazardous—containing flammable fumes, dust, etc. Some motors are waterproof, enabling them to operate under water.

The field poles (Fig. 8-1) are made from iron, either in solid form or built of thin strips called laminations. The iron field poles support the field windings and complete the magnetic circuit between the frame and armature core.

Bearings used in dc motors fit around the armature shaft and support the weight of the armature as shown in Fig. 8-2. They are made in three general types: sleeve, roller, and ball bearings.

Oil rings, also shown in Fig. 8-2, are small rings used with sleeve-type bearings. They carry the oil from the oil well to the shaft. The oil ring must turn when the machine is operating, or otherwise the bearing will burn out.

The rocker arm supports the brush holders. This arm is usually adjustable to make it possible to shift the brushes to obtain best operation. When the brushes are rigidly fastened to the end bell, the entire end-bell assembly is shifted to obtain the best possible operation. See Fig. 8-3.

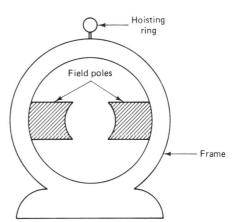

FIG. 8-1 Cross sections of a motor frame showing interior field poles.

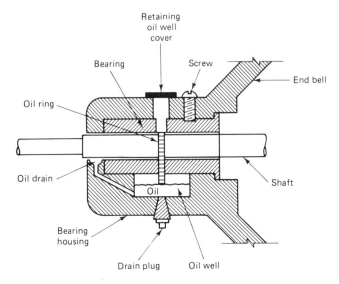

FIG. 8-2   Cross section of end bell showing bearing, oil well, and other details of construction.

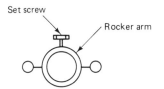

FIG. 8-3   Motor rocker arm.

The brush holders support the brushes and hold them in the proper position on the commutator. The brushes should be spaced equidistantly on the commutator when more than two sets of brushes are used. When only two sets are used, they will be spaced the same distance as a pair of adjacent field poles.

The brush tension spring applies enough pressure on the brush to make a good electrical connection between the commutator and brush. See Fig. 8-4.

Brushes used on dc motors are made of copper, graphite, carbon, or a mixture of these materials. The purpose of the brushes is to complete the electrical connection between the line circuit and the armature winding.

Commutators (Fig. 8-5) are constructed by placing copper bars or segments in a cylindrical form around the shaft. The copper bars are insulated from each other and from the shaft by mica insulation. An insulating compound is used instead of mica on small commutators. The commutator bars are soldered to and complete the connection between the armature coils.

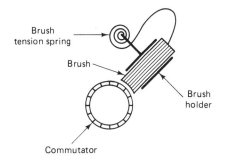

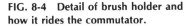

FIG. 8-4 Detail of brush holder and
how it rides the commutator.

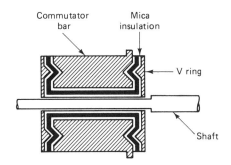

FIG. 8-5 Cross section of commutator.

The armature core (shown in Fig. 8-6) is made of laminated iron pressed tightly together. The laminated construction is used to prevent induced currents (eddy currents) from circulating in the iron core when the machine is in operation. The iron armature core is also a part of the magnetic circuit for the field and has a number of slots around its entire surface in which the armature coils are wound.

The armature winding (Fig. 8-7) is a series of coils wound in the armature slots, and the ends of the coils connect to the commutator bars. The number of turns and the size of wire are determined by the size, speed, and operating voltage of the machine. The purpose of the armature winding is to set up magnetic poles on the surface of the armature core.

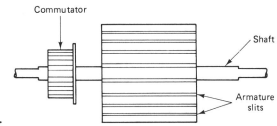

FIG. 8-6 Unwound motor armature.

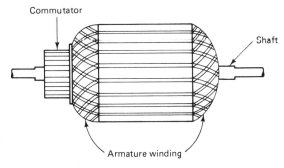

FIG. 8-7 Wound motor armature
showing commutator and shaft.

The field windings are made in three different types: shunt-, series-, and compound-wound fields. Shunt fields have many turns of small wire, and series fields have a few turns of heavy wire. The compound field is a combination of the two windings. The name of the field winding depends on the connection with respect to the armature winding, and its purpose is to produce magnetic poles that react with the armature poles to produce rotation.

SHUNT-WOUND DC MOTORS.   In this type of motor (Fig. 8-8), the strength of the field is not affected appreciably by change in the load, so that a relatively constant speed is obtainable. This type of motor may be used for the operation of machines requiring an approximate constant speed and imposing low starting torque and light overload on the motor.

The shunt-wound dc motor can be made into an adjustable-speed motor by means of field control or by armature control. For example, if a variable resistance is placed in the field circuit, the amount of current in the field windings along with the speed of rotation can be controlled. As the speed increases, the torque decreases proportionately, resulting in nearly constant horsepower. A speed range of 6 to 1 is possible using field control, but a lesser ratio is more common.

Speed regulation is somewhat greater than in the constant-speed shunt-wound motors, ranging from about 15 to 22%. If a variable resistance is placed in the armature circuit, the voltage applied to the armature can be reduced, in turn reducing the speed of rotation over a range of about 2 to 1. With the armature control, speed regulation becomes poorer as speed is decreased and is about 100% for a 2 to 1 range. The torque of this type of motor remains constant because the current through the field remains unchanged during operation.

Adjustable-speed shunt-wound motors are very useful on larger machines of the boring mill, lathe, and planer type and are particularly adapted to spindle drives because constant-horsepower characteristics permit heavy cuts at low speed and light or finishing cuts at high speed. They have long been used for planer drives because they can provide an adjustable low speed for the cutting stroke and a high speed for the return stroke. Their application has been limited, however, to plants where dc power is available, nowadays requiring the plant itself to either generate dc power or convert ac to dc.

SERIES-WOUND DC MOTORS.   In motors of this type (Fig. 8-9), any increase in load results in more current passing through the armature and the field windings. As the field is strengthened by this increased current, the motor speed decreases. Conversely, as the load is decreased, the field is weakened, and the speed increases, and at very light loads speed may become excessive. For this reason, series-wound motors are usually

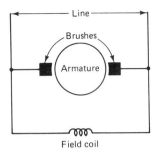

FIG. 8-8   Schematic of shunt-wound dc motor.

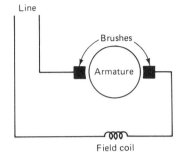

FIG. 8-9   Diagram of series-wound dc motor.

directly connected or geared to the load to prevent *runaway*. The increase in armature current with an increasing load produces increased torque, so that the series-wound motor is particularly suited to heavy starting duty and where severe overloads may be expected. Its speed may be adjusted by means of a variable resistance placed in series with the motor, but due to variation with load, the speed cannot be held at any constant value. This variation of speed with load becomes greater as the speed is reduced. Use of this motor is normally limited to traction and lifting service.

COMPOUND-WOUND MOTORS.   In this type of motor (Fig. 8-10), the speed variation—due to load changes—is much less than in the series-wound motor but greater than in the shunt-wound motor. It also has a greater starting torque than the shunt-wound motor and is able to withstand heavier overloads. However, it has a narrower adjustable-speed range. Standard motors of this type have a cumulative-compound winding, the differential-compound winding being limited to special applications. They are used where the starting load is very heavy or where the load changes suddenly and violently as with reciprocating pumps, printing presses, and punch presses.

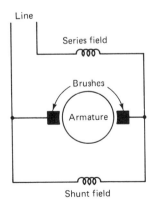

FIG. 8-10   Diagram of compound-wound motor.

Compound-wound motors use both shunt (parallel) and series field coils and have characteristics between both. In general, there are four different kinds of compound motors:

1. Long-shunt cumulative
2. Long-shunt differential
3. Short-shunt cumulative
4. Short-shunt differential

LONG-SHUNT CUMULATIVE.   In this type of motor, the current flows through the series field and shunt field coils of a pole in the same direction as shown in Fig. 8-11. When the shunt field is connected across the line, it is given the name of long shunt.

LONG-SHUNT DIFFERENTIAL.   If the shunt field connection of a compound motor is reversed with respect to the series field, the current will flow through it in the opposite direction as shown in Fig. 8-12. This produces bucking fields, and the motor is known as a differentially connected motor. Therefore, a motor with its shunt field connected across the line so that the series and the shunt fields have opposite polarity in the same pole is known as the long-shunt differential motor.

SHORT-SHUNT CUMULATIVE.   When the shunt field of a compound motor is connected to the armature terminals instead of across the line, the motor is known as a short-shunt motor. When the shunt is connected across the armature so that the current flows through it in the

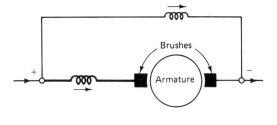

FIG. 8-11   Current flowing through the series field and shuntfield coils of a pole in the same direction.

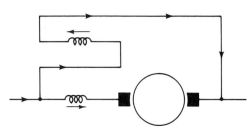

FIG. 8-12   Current flowing in the opposite direction through shunt and series coils.

same direction as in the series field, the motor is known as a short-shunt cumulative motor.

SHORT-SHUNT DIFFERENTIAL. If connected so that the current flows through the shunt field in the opposite direction to the current in the series field, the motor is known as a short-shunt differential motor as shown in Fig. 8-13.

Compound motors, like series motors, are known as variable-speed motors because the speed varies with the load. Therefore, the compound motor is normally used for two types of work: loads requiring relatively constant speed but having frequent peaks needing high torque such as compressors and reciprocating pumps and on applications requiring the motor to sometimes run without load, such as winches, hoists, and the like.

High starting current must be reduced by adding external resistance to the motor; the speed is regulated by changing the armature resistance, the field strength, or the applied voltage. Motor ratings depend on temperature limitations during operation, and manufacturers provide guarantees for allowable speed regulation of different types of motors.

BRUSHLESS DC MOTORS

The brushless dc motor was developed to eliminate commutator problems in missiles and spacecraft in operation above the earth's atmosphere. Two general types of brushless motors are in use: the inverter-induction motor and a dc motor with an electronic commutator.

The inverter-induction motor uses an inverter which uses the motor windings as the usual filter. The operation is square wave, and the combined efficiencies of the inverter and induction motor are at least as high as for a dc motor alone. In all cases, the motors must be designed for low starting current, or else the inverter must be designed to saturate so that starting current is limited; otherwise the transistors or silicon-controlled rectifiers in the inverter will be overloaded.

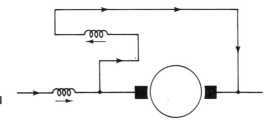

FIG. 8-13 Short-shunt differential motor.

MOTOR BRAKING

Motor brakes are composed of a linkage of moving parts and a friction member. Different types of brakes are available depending on the power supply, ratings, and the application requirements.

The torque ratings of the brake is usually selected at least equal to the full-load torque of the motor. There are cases in which this value should be exceeded, where it is necessary to stop the machine very quickly or to provide an adequate margin of safety.

The time rating of brakes is controlled by the amount of heating produced in the brake coil or thrustor motor. The brake time rating is usually made at least equal to the motor time rating, since the brake is energized as long as the motor is energized.

Continuous rated brakes are applied where conditions of intermittent-duty rating cannot be met. Also continuous rated brakes should be applied when any doubt exists that the intermittent-duty rating will be exceeded.

Intermittent-duty shunt-wound dc brakes should not be energized for more than one half of the *off-on* cycle with continuous application of normal voltage not exceeding 1 h. The same intermittent-duty time applies for ac brake coils and thrustor brakes with the exception that they shall not release more than three times per minute.

Since dc series brake coils are applied only with series-wound motors – which are usually applied either on a ½- or 1-h basis – the brake coil time rating is again made at least equal to the motor time rating. Disc brakes are usually applied on the basis of continuous duty.

Types SA and SAR self-adjusting dc magnetic brakes can be used for fast stopping of any dc motored drive. They are self-adjusting, require minimum maintenance, and are supplied with either shunt or series coils.

Direct-current-type SAR Rectox operated brakes are employed where the smooth operation of a dc brake is desired on an ac motor. The SAR brake uses a standard shunt coil plus an interlock. This brake is capable of applying the intermittent torque rating continuously due to the interlock feature.

DYNAMIC BRAKING

Dynamic braking can be used with synchronous and induction motors. On dc shunt and compound motors, the armature is disconnected from the line and quickly connected to a resistor. The shunt field remains energized. The generator or braking action decreases with the speed and stops the motor quickly. Most machines, especially machine tools, have considerable friction, and dynamic braking provides extremely rapid and uniform results with no mechanical wear or maintenance. However, a machine with little friction would drift at lower speeds. Also, with such a machine, an overhaulting load might not stop with the motor.

On adjustable-speed motors, a relay is provided which short-circuits the field theostat during braking, giving maximum field strength and materially decreasing the time of travel during braking. Schematic braking connections for shunt motors, of either constant or adjustable speed, are shown in Fig. 8-13.

Dynamic braking is used on synchronous motors for quick stopping when a safety bar is operated in the case of emergency. The dc field is left energized, and the motor is disconnected from the ac line and connected to a three-phase resistor. By using quick-acting control sequences and the proper resistors, any size motor can be stopped in approximately 1 s without undue shock or stresses.

Dynamic braking of induction motors is comparatively new, but it is growing rapidly in applications for many textile and machine tool drives. When the starter is opened, dc voltage is applied to two or more stator terminals. Suitable interlocking of contactors is required so that the starter and dc contactor cannot be closed at the same time; also, a timing relay is required to open the braking contactor after the motor has stopped. The resistance of ac motor windings is low, so that if a 125-V dc supply is used, external resistors are generally required to limit the braking current. As a rule, satisfactory braking is obtained with a current from two to three times the full-load rated amperes of the motor.

With dynamic braking, there is no tendency to reverse the motor, and where safety to operators is involved, this feature is often of major importance.

9

Direct-Current Motor Control

Direct-current controllers are classified as many different types, but essentially they are either manually or automatically operated.

DC CONTROLLERS

Small dc motors of, say, less than ½ hp consume very little current upon starting and therefore can be started by placing full voltage across the motor terminals. On the other hand, large dc motors cause large initial currents to flow because they have a low resistance, and the excessive current flow during starting may damage the motor or trip the overcurrent device.

To start a large dc motor, it is necessary to place a resistance unit in series with the motor so that the starting current is reduced to a safe value. As the motor accelerates, this resistance can be gradually decreased. Once the motor reaches the desired speed, the resistance is no longer necessary because the motor is now generating a voltage which is in opposition to the impressed voltage, thereby preventing excessive current flow. This opposing voltage is called the counterelectromotive force (CEMF), and its value will depend on the speed of the motor, which is greatest at full speed and zero at standstill.

In general, dc motors of less than 2-hp rating can be started across the line, but with larger motors it is usually necessary to put resistance in series with the armature when it is connected to the line, as discussed previously. This resistance, which reduces the initial starting current to a point where the motor can commutate successfully, is shorted out in steps as the motor comes up to speed and the countervoltage generated is sufficient to limit the current peaks to a suitable value. Accelerating contactors that short out successive steps of starting resistance may be controlled by countervoltage or by definite-time relays.

For small motors used on auxillary devices the CEMF starter is satisfactory. The definite-time starter is more widely used, however, and has the advantage of being independent of load conditions.

The following diagrams illustrate some of the circuits commonly used for dc motor control.

Basic requirements of a nonreversing dc starter in its simplest form is shown in Fig. 9-1. When the start push button is depressed, line contactor M closes, energizing the motor armature through the starting resistance. As the motor comes up to speed, the countervoltage, and the voltage across the motor armature and series field, increases. At a predetermined value the accelerating contactor A closes, shorting out the starting resistance.

Figure 9-2 shows a typical nonreversing, constant-speed, definite-time starter. The accelerating contactor is equipped with a time delay mechanism. This contactor A is of the magnetic-flux-decay type. It is spring-closed, equipped with two coils, and has a magnetic circuit that retains enough magnetism to hold the contactor armature closed and the contact open indefinitely. Main coil Am has sufficient pull to pick up the armature and produce permanent magnetization. Neutralizing coil An is connected for polarity opposite the main coil. It is not strong enough to affect the pickup or holding ability of the main coil, but when the latter is deenergized, the neutralizing coil will buck the residual magnetism so

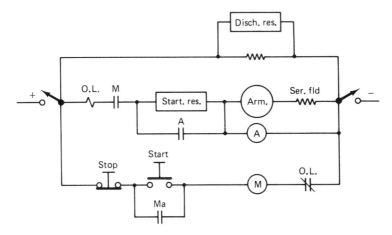

FIG. 9-1   Basic requirements of a nonreversing dc motor starter.

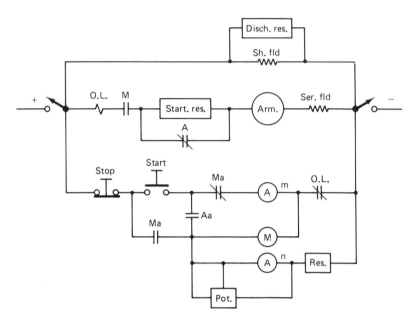

FIG. 9-2   Diagram for nonreversing, constant-speed, definite-time dc starter.

that the contactor armature is released by the spring, and the contacts close. By adjusting the potentiometer, the voltage impressed on this coil and hence the time required for the contactor to drop out can be varied. When the start button is depressed, accelerating contactor coil Am is energized, causing contact A to open and auxiliary contact Aa to close. Contact Aa energizes line contactor M, and normally open auxiliary contacts Ma establish a holding circuit. Neutralizing coil An is also energized. Opening of contact Ma deenergizes coil Am, and contactor A starts timing. At the set time the main normally closed contacts on A close, shorting out the starting resistance and putting the motor across the line.

The same kind of a starter as in Fig. 9-2 but designed for use with a motor of larger horsepower is shown in Fig. 9-3. This starter provides two steps of definite-time starting. The operation is essentially the same as in Fig. 9-2, but the first accelerating contactor $1A$ does not short out all the starting resistance. It also starts $2A$ timing, which finally shorts out the remaining resistance. The normally open auxiliary contacts on the accelerating contactors in Figs. 9-2 and 9-3 are arranged so that it is necessary for the accelerators to pick up before the line contactor can be energized. This is a safety interlocking scheme that prevents starting the motor across the line if the accelerating contactors are not functioning properly.

One way of producing dynamic braking is shown in Fig. 9-4. Control circuits have been omitted, since they are a duplicate of those shown in

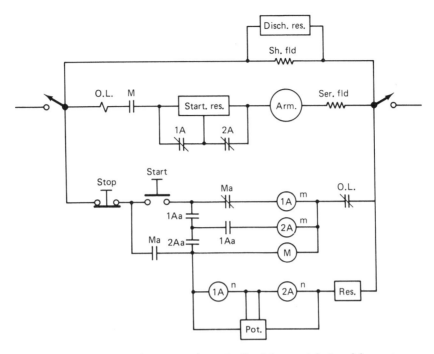

FIG. 9-3 Same type of starter as shown in Fig. 9-2 except designed for motors with larger horsepower.

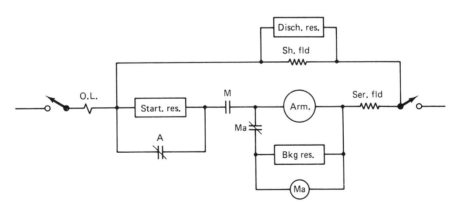

FIG. 9-4 One method of producing dynamic braking on a dc motor.

Figs. 9-2 and 9-3. Line contactor M has two poles, one normally open and the other normally closed. Both poles are equipped with an operating coil and are on the same armature, which is hinged between the contacts. In starting, when line contactor M closes, normally closed contact Ma opens. When the stop button is depressed, the line contactor drops out, and contact *Ma* closes. The motor, now acting as a generator, is connected to the braking resistor, and coil *Ma* is energized by the resultant voltage. It causes *M* to seal in tightly, establishing good contact pressure and preventing this contact from bouncing open.

In the more modern types of controllers a separate spring-closed contactor is used for dynamic braking. See Fig. 9-5. Operation is similar to that described for Fig. 9-2, except that the energizing of coil *Am* and the picking up of accelerating contactor *A*, closing contact *Aa*, energize dynamic braking contactor *DB*, which in turn energizes line contactor *M* through its auxiliary contact *DBa*. This arrangement not only ensures that the dynamic braking contactor is open but also that it is open before the line contactor can close. To obtain accurate inching, such as is required for most machine tool drives, the motor must respond instantly to the operation of the push button. In the scheme shown in Fig. 9-5, the closing of the line contactor is delayed until the accelerating contactor and the dynamic braking contactor pick up.

Figure 9-6 shows an arrangement to secure quicker response of the motor for more accurate inching. Accelerating contactors 1*A* and 2*A* are energized in the off position. Hence, when the start button is depressed, the dynamic braking contactor picks up immediately, and its auxiliary contact *DBa* picks up *M* line contactor.

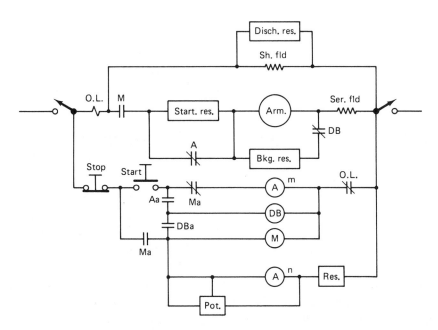

FIG. 9-5 Separate spring-closed contactors are used in some applications of dynamic braking.

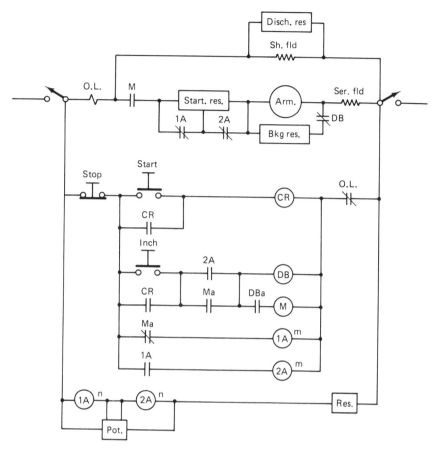

FIG. 9-6 Control for inching a dc motor.

One method of connecting the full-field relay, used with adjustable-speed motors having a speed range in excess of 2 to 1, is shown in Fig. 9-7. Coil *FF* is energized by the closing of the normally open auxiliary contact *Aa* and remains closed until the last accelerating contactor drops out. Contacts of the full-field relay *FF* are connected to short out the field rheostat, thereby applying maximum field strength to the motor during the starting period.

Another method of applying the full-field relay is shown in Fig. 9-8. This arrangement ensures full field on starting and provides for limiting the armature current when the motor is accelerating from the full-field speed to the speed set by the rheostat. Field accelerating relay *FA* is equipped with two coils, one a voltage coil connected across the starting resistance and the other a current coil connected in series with the motor armature. See Fig. 9-2 for the remainder of the circuit. When line contac-

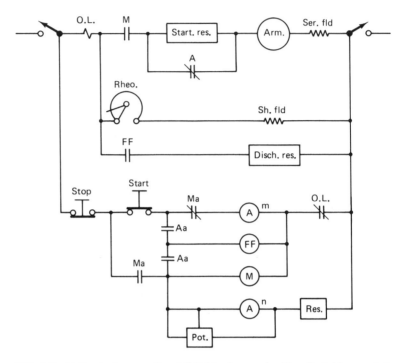

FIG. 9-7 Method of connecting full-field relay used with adjustable-speed dc motors.

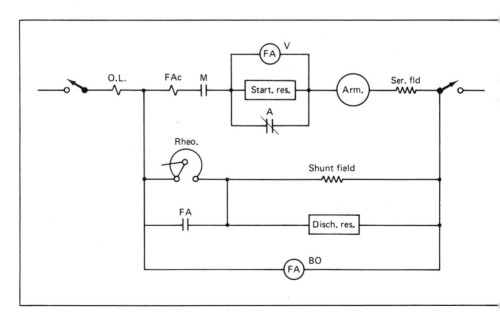

FIG. 9-8 This wiring arrangement ensures full field on starting.

tor *M* closes, the voltage drop across the starting resistor is practically line voltage, and relay *FA* is picked up quickly. When accelerating contactor *A* closes, voltage coil *FAv* is shorted, but the closing of *A* produces a second current peak, and current coil *FAc* holds relay *FA* closed. As the motor approaches full-field speed, this current decays and allows the *FA* contacts to open, weakening the motor field. When the motor attempts to accelerate, the line current again increases. If it exceeds the pickup value of coil *FAc*, the relay will close its contacts, arresting acceleration and causing a decay of line current, which again causes *FA* to drop out. High inductance of the motor field plus inertia of the motor and drive prevent rapid changes in speed. Hence the motor will not reduce its speed, but the increased field current will reduce the armature current and cause *FA* to drop out. The fluttering action will continue until the motor reaches the speed set by the rheostat. Setting of the *FA* relay current coil determines the maximum current draw during this part of the acceleration period. Since relay *FA* must handle the highly inductive field circuit, a good blowout arrangement is necessary. Hence the relay is usually equipped with a shunt blowout coil, *FABO*.

Connections of field lost relay, to prevent excessive speed if the shunt field is deenergized while voltage remains on the armature, is shown in Fig. 9-9. It usually consists of a current relay in series with the

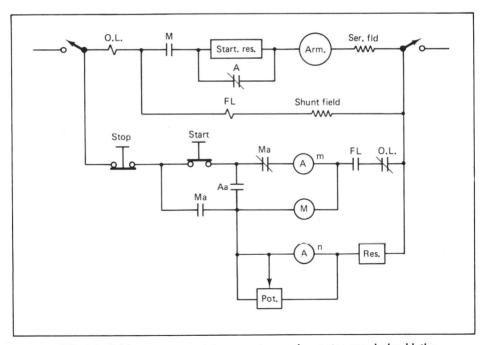

FIG. 9-9 Wiring arrangement to prevent excessive motor speed should the shunt field be deenergized while voltage remains on the armature.

motor shunt field and is adjusted to pick up on full-field current and remain closed at any current within the operating range of the motor field current. Contacts of relay *FL* are connected in series with the overload relay contacts so that the opening of its contacts will deenergize the control by opening the line contactor. This type of field loss protection does not protect against the possibility of a short circuit across a part of the field, say across the one field coil. This would cause the motor speed to rise considerably, but the current in the field circuit would also rise. Consequently, the series current relay would not respond.

VARIABLE-VOLTAGE CONTROL

The variation in speed obtainable by field control on the conventional dc motor will not, in most cases, exceed 4 to 1 due to the sparking difficulties experienced with very weak fields. Although the range may be increased by inserting resistance in series with the armature, this can be done only at the expense of efficiency and speed regulation.

With constant voltage applied to the field, the speed of a dc motor varies directly with the armature voltage; therefore, such a motor may be steplessly varied from zero to maximum operating speed by increasing the voltage applied to its armature. The diagram in Fig. 9-10 shows the arrangement of machines and the connections used in one type of variable-voltage control designed to change speed and reverse rotation. The constant-speed dc generator (*B*) is usually driven by an ac motor (*A*), and its voltage is controlled by means of a rheostat (*R*). Note that the

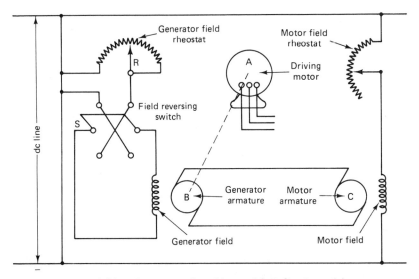

FIG. 9-10   Variable-voltage control used in special applications of dc motors.

fields of both generator and driving motor are energized from a separate dc supply or by an auxiliary exciter driven off the generator shaft, causing the strength of the motor field to be held constant, while the generator field may be varied widely by rheostat R.

With the system in operation, generator B is driven at a constant speed by prime mover A. Voltage from B is applied to the dc motor (C) which is connected to the machine to be driven. By proper manipulation of rheostat R and field reversing switch S, the dc motor may be gradually started, brought up to and held at any speed, or reversed. As all of these changes may be accomplished without breaking lines to the main motor, the control mechanism is small, relatively inexpensive, and less likely to give trouble than the equipment designed for heavier currents.

The advantage of this system is the flexibility of the control and also the complete elimination of resistor losses, the relatively great range over which the speed can be varied, the excellent speed regulation on each setting, and the fact that changing the armature voltage does not diminish the maximum torque which the motor is capable of exerting, since the field flux is constant.

By means of the arrangement shown in Fig. 9-10, speed ranges of 20 to 1 – as compared to 4 to 1 for shunt field control – may be secured. Speeds above the rated normal full-load speed may be obtained by inserting resistance in the motor shunt field. This represents a modification of the variable-voltage control method which was originally designed for the operation of constant-torque loads up to the rated normal full-load speed.

As three machines are usually required, this type of speed control finds application only where great variations in speed and unusually smooth control are desired. Steel mill tools, electric shovels, passenger elevators, machine tools, turntables, large ventilating fans, and similar equipment represent the type of machinery to which this method of speed control has been applied.

OVERLOAD RELAYS

To protect the motor and related circuits from accidental or prolonged overloads, either the starter or the motor should be equipped with automatic devices that will open the circuit should an overload exist. This protection can be provided by fuses, circuit breakers, or overload relays.

Overcurrent protection must be provided in the line of every motor circuit, but additional protection should be provided in the form of magnetic overload relays. These are used in both manual and automatic starters.

Figure 9-11 shows a plunger type of overload relay. When the current through the coil reaches the value set by the adjustable screw, the

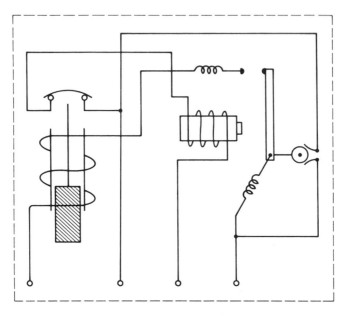

FIG. 9-11   Plunger type of overload relay.

plunger is drawn up and opens two contacts. This type of relay can be used on both the manual and automatic controller.

Most overload relays used on modern starters are thermally operated and usually consist of two strips of metal having different degrees of thermal expansion which are welded together. If this bimetallic strip is heated, it will deflect sufficiently to trip two normally closed contacts, which in turn will open-circuit the holding coil of a magnetic contactor, causing the main contacts to open. An advantage of this protective device is that it provides a time delay which prevents the circuit from being opened by momentary high starting currents and short overloads. At the same time it protects the motor from prolonged overloading.

More information on overload relays may be found in Chapter 6.

10

Universal, Shaded Pole, and Fan Motors

The three types of motors discussed in this chapter are quite common and are used for a variety of applications – especially small appliances. All are single-phase motors, and most are designed to operate either on 120 or 240 V.

UNIVERSAL MOTORS

The universal motor is a special adaptation of the series-connected dc motor, and it gets its name *universal* from the fact that it can be connected on either ac or dc and operate the same. All are single-phase motors for use on 120 or 240 V.

In general, the universal motor contains field windings on the stator within the frame, an armature with the ends of its windings brought out to a commutator at one end, and carbon brushes which are held in place by the motor's end plate, allowing them to have a proper contact with the commutator.

When current is applied to a universal motor, either ac or dc, the current flows through the field coils and the armature windings in series. The magnetic field set up by the field coils in the stator reacts with the current-carrying wires on the armature to produce rotation.

You will find universal motors used on such household appliances as sewing machines, vacuum cleaners, electric fans, and the like.

The universal motor shown in Fig. 10-1 operates on the magnetic

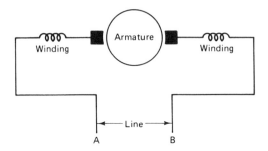

FIG. 10-1   Wiring diagram of a universal motor.

interaction between the armature and field poles and runs in the same direction whether the current flows in on line A or on line B, since reversing the flow of current in the line wires changes the polarity of both the armature and field poles at the same instant. Therefore, if such a motor is supplied with alternating current, the torque developed will always be in the same direction. However, to operate successfully on ac, all parts of the magnetic circuit must be laminated to prevent undue heating from eddy currents, and element windings are usually desirable on the armature to ensure acceptable commutation. On the compensating windings of larger motors improvements have been made to provide more efficient operation and to reduce brush sparking.

The universal motor will produce about four times normal full-load torque with two times normal full-load current. The torque produced increases very rapidly with an increase in current. The variation in speed from no load to full load is so great that complete removal of the load is dangerous in all motors of this type with the exception of fractional horsepower ones.

This type of motor is widely used in fractional hp sizes for fans, vacuum cleaners, kitchen mixers, and portable equipment of all types such as electric drills, sanders, saws, etc. Higher ratings are employed in traction work and for cranes, hoists, and so on. In general, universal motors are suitable for applications where high starting torque or universal operation is desired.

Principal troubles occur due to wear of the commutator, brushes, brush holders, or bearings; opens, shorts, or grounds occur in the armature, field, or associated apparatus; loose connections are another source of trouble. To reverse the direction of rotation, reverse the armature connections or the field connections but not both.

SHADED POLE MOTOR

A shaded pole motor is a single-phase induction motor provided with an uninsulated and permanently short-circuited auxiliary winding displaced in magnetic position from the main winding. The auxiliary winding is

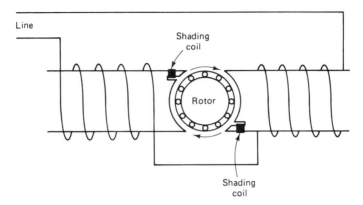

Line

Shading coil

Rotor

Shading coil

FIG. 10-2 Wiring diagram of a shaded pole motor.

known as the shading coil and usually surrounds from one third to one half of the pole. See Fig. 10-2. The main winding surrounds the entire pole and may consist of one or more coils per pole.

In the unshaded section of the pole the magnetic flux produced by the main winding is in phase with the main winding current, whereas the flux produced by the shading coil acts as a phase-splitting device to produce the rotating field that is essential to the self-starting of all straight induction motors. As the movement of the flux across the pole face is always from the unshaded to the shaded section of the pole, the direction of rotation can be determined on the normally nonreversible motor by noting the position of the shading coil with respect to the pole itself. This type can be reversed by removing the stator from the frame, turning it through 180°, and replacing it.

The starting torque will not exceed 80% of full-load torque at the instant of starting, increases to 120% at 90% of full speed, and decreases to normal at normal speed. This type of motor operates at low efficiency and is constructed in sizes generally not exceeding $\frac{1}{20}$ hp.

Applications for this motor include small fans, timing devices, relays, radio dials, or any constant-speed load not requiring high starting torque.

DISTRIBUTED-FIELD COMPENSATED MOTOR

The distributed-field compensated motor, a type of universal motor, has a stator coil similar to that of the split-phase motor, and two types are normally available: the single-field compensated motor and the two-field compensated motor. The former has one stator winding, while the latter has two.

The two-pole compensated motor has a stator winding like the main winding of a two-pole split-phase motor, with fields wound into the slots of the stator. Field poles, of course, must be of opposite polarity and connected in series with the armature. This type of motor may be reversed by interchanging either the armature or field leads and then shifting the brushes against the direction in which the motor will rotate.

Two windings are used in the stator of the two-field compensated motor, similar to the starting and running windings of a split-phase motor; they are located 90 electrical deg from one another. The compensating winding is used to reduce the reactance voltage present in the armature when it is operating on alternating current, caused by the alternating flux.

The speed control of this type of motor may be regulated by several methods, some of which include a centrifugal switch, using a tapped field, or inserting resistance in series with the motor.

REVERSING SHADED POLE MOTORS

Some shaded pole motors are designed to be reversed by means of a built-in control switch, but most motors of this type require internal changes before they can be reversed. One method is to disassemble the motor, reverse the stator end for end, and then reassemble.

On motors of this type that can be reversed externally, one main winding is present along with two separate shaded windings. The stator of this motor has slots into which the windings are placed. The main winding is usually distributed over several slots but may have only one coil per pole. To reverse the motor, the closed shaded winding is opened, while the other shaded winding is closed, changing the position of the shaded poles with reference to the main poles.

11

Generators

In very general terms, a motor and a generator may be said to be opposites; that is, a motor is a machine that when supplied with electric current can be used for mechanical work, such as driving an industrial machine and the like. A generator, on the other hand, is a machine that is driven by mechanical means to produce electric current. Most generators are rated in kilowatts (kW) and range in size from a fraction of a kilowatt to several thousand kilowatts. A typical generator is shown in Fig. 11-1. This type is portable and is suitable for supplying 120/240 ac up to 3 kW.

DIRECT-CURRENT GENERATORS

Direct-current generators are very similar to dc motors in both appearance and construction. Both have the same types of armature and field poles and are generally identical. For this reason, a dc motor can easily be converted into a generator or vice versa.

If a piece of wire is moved through magnetic lines of force, as shown in Fig. 11-2, so the wire cuts across the path of the flux, a voltage will be induced in the wire. Faraday first made this discovery in about 1831. If we then connect a sensitive meter to this wire – thus completing the circuit – the needle will indicate a flow of current every time the wire is moved across the lines of force. This induction, of course, generates voltage in the wire, without producing any current, if the circuit is open. In fact, current is never generated; rather, pressure is set up which causes

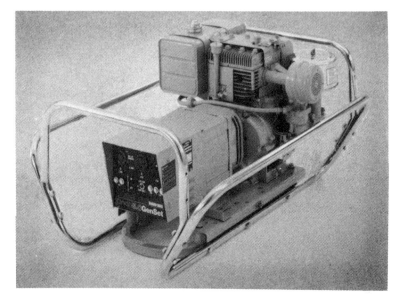

FIG. 11-1

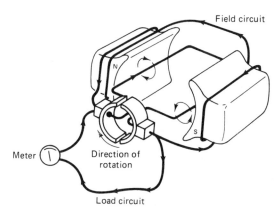

Meter ⓘ   Direction of
                    rotation   **FIG. 11-2**

Load circuit

current flow if the circuit is completed. So current always results from the production of pressure first and only when the circuit is closed.

A simple generator is shown in Fig. 11-3 which is a single-turn coil arranged to be revolved in the field of permanent magnets. The ends of the coils are attached to metal slip rings which are fastened to the shaft and revolving with it. This arrangement gives a connection from the moving coils to the circuit load by means of metal or carbon brushes rubbing on the slip rings.

Assume that the coil in Fig. 11-3 revolves to the right, or clockwise.

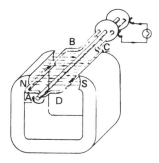 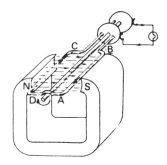

FIG. 11-3 [Reprinted with permission from John Traister, *Handbook of Power Generation: Transformers and Generators* (Englewood Cliffs, N.J.: Prentice-Hall, Inc. © 1983), p. 2.]

FIG. 11-4 [Reprinted with permission from John Traister, *Handbook of Power Generation: Transformers and Generators* (Englewood Cliffs, N.J.: Prentice-Hall, Inc. © 1983), p. 2.]

Wire AB will be moving upward through the flux, and the induced pressure will be in the direction indicated by the arrow on it. At the same time, wire CD is moving downward, and its induced pressure will be in the reverse direction but will join with and add to that of wire AB, as they are connected in series in the loop. Note that the current flows to the nearest collector ring and out along the lower wire to the load, returning on the upper wire to the farthest collector ring and the coil.

Figure 11-4 shows the same coil in Fig. 11-3 after it has turned one-half revolution farther, and now wire AB is moving downward instead of up as before. Therefore, its pressure and current are reversed. Wire CD is now in the position where AB was before, and its pressure is also reversed. This time the current flows out to the farthest collector ring and over the top wire to the load, returning on the lower wire.

The discussion thus far has been that of an alternator. If direct current is desired, a commutator must be used, or else some other type of rotary switch, to reverse the coil leads to the brushes as the coil moves around. All common generators produce alternating current in their windings, so we must convert it in this manner if direct current is desired in the external circuit.

In Fig. 11-5, another revolving loop is shown. Note that wire AB is moving upward, that its current is flowing away from the front of the coil, and that the current of wire CD is flowing in the opposite direction. The coil ends are connected to two bars or segments of a simple commutator, each wire to its own separate bar. With the coil in this position, the current flows out at the right-hand brush, through the load to the left, and reenters the coil at the left brush.

In Fig. 11-6, the coil has moved one-half turn to the right, and wire AB is now moving down, and its current is reversed. However, the commutator bar to which it is connected has also moved around with the

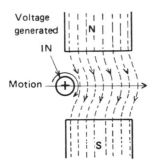

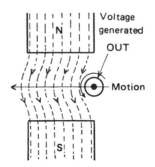

FIG. 11-5   (Courtesy Page Power Co.)   FIG. 11-6   (Courtesy Page Power Co.)

wire, so the current still flows in the same direction in the external circuit, producing direct current to the load.

In Fig. 11-6 is shown another illustration of power generation. At *A*, the lines of force from the field poles are passing downward, and the conductor is being moved to the right. This will induce a voltage in the wire that will tend to cause current to flow in at the end which is visible. In *B* in Fig. 11-6, the conductor is moving in the opposite direction, through the same magnetic field; note that the induced voltage has reversed with the direction of the conductor movement.

The circular arrows around the conductors in Fig. 11-6 indicate the direction of the lines of force which will be set up around them by their induced currents.

Refer to Fig. 11-7, which shows two conductors of a coil mounted in slots of an armature and revolving clockwise. In their position at *A*, the

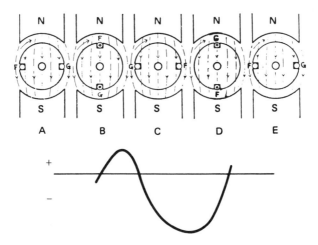

FIG. 11-7   (Courtesy Page Power Co.)

conductors are not generating any voltage, as they are in the neutral plane and are not cutting across lines of force. At *B*, the direction of induced voltage will be *in* at conductor *F* and *out* at *G*; if the conductors are connected together at the back of the coil, their voltages will add together.

In position *C*, the conductors are both in the neutral plane again, so their induced voltage once more falls to zero.

At position *D*, conductor *G* is passing the north pole and conductor *F* is passing the south pole, so they are both moving through the field flux in opposite directions to what they were at position *B*, and therefore their induced voltage will be reversed. At *E*, both conductors are again back in the neutral plane, or at the same point they started from.

A curve indicating the voltage generated is shown below these various steps of generation in Fig. 11-7. At *A*, the voltage curve is starting at the zero line, as the conductors start to enter the field flux. At *B*, where the conductors are cutting through the dense field directly under the poles, the curve shows maximum positive voltage. From this point, it falls off gradually as the conductors pass out of the flux at the poles until it again reaches zero at *C*. Then as the conductors each start to cut flux in the opposite direction, the curve shows negative voltage in the opposite direction or below the line, reaching maximum value at *D*. At *E*, the negative voltage has again fallen to zero.

CYCLES AND ALTERNATIONS

When one complete revolution is completed in the simple two-pole generator, it also completes what is called one cycle of generated voltage. Referring again to Fig. 11-7, note the single positive impulse produced by the conductor passing one complete pole, as indicated by the curve from *A* to *C*. This is called one alternation, two of which are required to make one cycle. Therefore, each time a conductor passes one north pole and one south pole, it produces one cycle.

There are 360° in a circle, or, for that matter, in one revolution of a conductor on an armature. When referring to generators, the term is usually changed and expressed in electrical degrees. Therefore, the conductor travels 360 electrical deg each time it passes two alternate field poles and completes one cycle. From this it is easily seen that one cycle consists of 360 electrical deg and one alternation consists of 180 electrical deg.

In generators with more than two poles, it is not necessary for the conductor to make a complete revolution to complete a cycle, since one cycle is produced for each pair of poles passed. So a four-pole generator, for example, would produce two cycles per revolution; an eight-pole generator, four cycles per revolution, and so forth.

FREQUENCY

Alternating-current circuits have their frequency expressed in cycles per second, the most common being 60 cps, or 60 Hz. Since frequency is expressed in cycles per second and a conductor must pass one pair of poles to produce a cycle, the frequency of an ac generator depends on the number of its poles and the speed of rotation.

For example, a four-pole generator rotated at 1800 rpm would produce a frequency of 60 cps since its conductors will pass two pairs of poles per revolution, or $1800 \times 2 = 3600$ pairs of poles per minute. Then, as there are 60 s in a minute, $3600/60 = 60$ cps.

A generator with, say, 12 poles would only need to rotate at 600 rpm to produce 60 cps. The conductor in such a generator would pass six pairs of poles per revolution or at 600 rpm would pass $6 \times 600 = 3600$ pairs of poles per minute, and $3600/60 = 60$ cps.

ALTERNATORS

Alternating-current generators are commonly called alternators, most of which are of the revolving field type; that is, the armature conductors are stationary while the field revolves – just the opposite of the generators discussed previously. This type of construction has two very important advantages for large power plant alternators. The first is if the armature conductors are stationary, the line wires can be permanently connected to them, and it is not necessary to take the generated current out through brushes or sliding contacts. This is quite an advantage with the heavy currents and high voltages produced by modern alternators.

Of course, it is necessary to supply the current to the revolving field with slip rings and brushes, but this field energy is many times smaller in amperes and lower in volts than the main armature current.

The other big advantage is that the armature conductors are much larger and heavier than those of the field coils and much more difficult to insulate because of their very high voltage. It is, therefore, much easier to build the armature conductors into a stationary element than it is in a rotating one.

The field, being the lighter and smaller element, is also easier to rotate, and this reduces bearing friction and troubles as well as air friction at high speeds.

With large revolving field alternators, the stationary armature is commonly called the stator, and the rotating field is called the rotor.

SINGLE-PHASE SYSTEMS

Figure 11-8 shows a simple revolving field alternator with one coil in the slots of the stator, or stationary armature. The circles in the slots show the ends of the coil sides, and the dotted portion is the connection between them at the back end of the stator. Inside the stator core is a two-pole field core with its coil mounted on a shaft so it can be revolved.

When direct current is supplied to the field core through the slip rings and brushes shown, the core becomes a powerful electromagnet with flux extending from its poles into the stator core. Then, as the field is revolved, the lines of force from its poles revolve with them and cut across the conductors in the stator slots.

As each coil side is passed first by the flux of the north pole and then a south pole, the induced voltage and current will be alternating, as it was with the revolving armature type previously shown. The sine wave beneath the generator shows the complete cycle which will be produced by one revolution of the two-pole field. This particular generator would have to revolve at 3600 rpm to produce 60-cycle energy. Such high-speed generators are normally used only for standby power and are never run continuously due to the wear on the parts at the high speed. Revolving fields

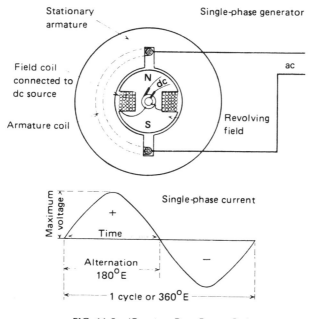

FIG. 11-8   (Courtesy Page Power Co.)

are made with four or more poles to produce 60-cycle energy at lower speeds.

TWO-PHASE SYSTEMS

Generators and alternators are also made to produce polyphases, such as two-phase and three-phase currents, meaning their currents are divided into more than one part.

Figure 11-9 shows a simple two-phase alternator, which has two separate coils placed in its stator at right angles to each other, or displaced 90° from each other. As the field of this generator revolves, it will induce voltage impulses in each of these coils, but these impulses will not come at the same time, due to the position of the coils. Instead, the voltages will be generated 90 electrical deg apart, as shown in the sine wave in Fig. 11-9.

Referring again to the curve in Fig. 11-9, curve A shows the voltage generated in coil A as the poles pass its sides. As these poles rotate 90° farther, their flux cuts across coil B and produces the voltage impulses shown by curve B which are all 90° later than those in curve A. These two separate sets of impulses are each carried by their own two-wire circuits as shown in the diagram.

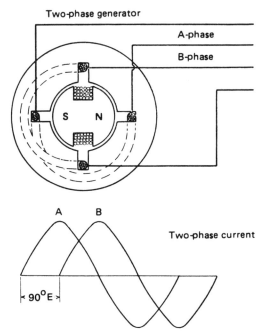

FIG. 11-9   (Courtesy Page Power Co.)

Consequently, a two-phase circuit is simply a circuit of two parts — or having two sets of alternations occuring 90° apart. Note from the curve in Fig. 11-9 that these alternations or impulses overlap each other and that while one is at zero value, the other is at maximum value, always maintaining voltage in one phase or the other as long as the circuit is *alive*. This feature is quite an advantage where the energy is used for power purposes, as these overlapping impulses produce a stronger and steadier torque than single-phase impulses.

For this same reason, three-phase energy is still more desirable for motor operation and power transmission and is much more generally used than two-phase energy.

THREE-PHASE SYSTEMS

The drawing in Fig. 11-10 shows a simple three-phase alternator, with three coils in its stator spaced 120 electrical deg apart. As the field poles revolve past coils *A*, *B*, and *C* in succession, they induce voltage impulses which are also 120° apart, as shown in the curves in the drawing.

The line leads are taken from the coils at points 120° apart, and the other ends of the coils are connected together at *F*. This type of connection is known as the star or Y connection of the coils to the line. Another

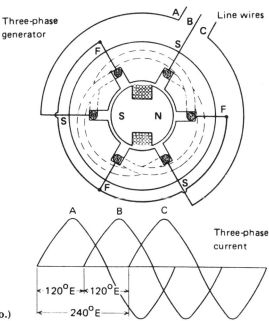

FIG. 11-10 (Courtesy Page Power Co.)

common connection for three-phase windings is the delta connection. See Figs. 11-11 and 11-12.

Figure 11-11 shows the method of making a star or Y connection with ac windings, while Fig. 11-12 shows delta connections. By comparing these two forms of connections, you will note that the delta connection has only half as many turns as the Y connection. Since the number of turns or coils in series directly affects the voltage, the Y connection will produce higher voltage than the delta, and when used on a motor, the Y connection will enable the motor to operate on higher line voltage.

The delta connection, however, has two windings in parallel between any two line or phase leads, so it will have greater line current capacity than the Y or star connection.

Since the star or Y connection places twice as many coils in series between the line wires as the delta connection, it might appear to give dou-

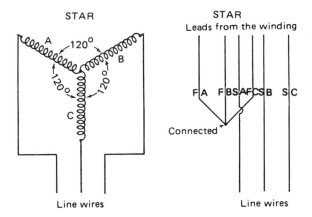

FIG. 11-11   (Courtesy Page Power Co.)

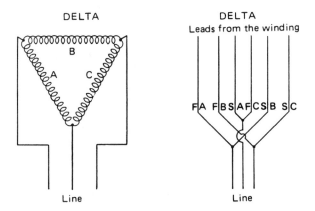

FIG. 11-12   (Courtesy Page Power Co.)

ble the voltage of a delta system. However, the voltage increase will not be double because the spacing of the two windings in the alternator is 120° apart, and consequently their maximum voltages occur at slightly different periods of time. The placing of the *C* phase winding between the windings of the *A* and *B* phases actually reverses its phase relation to the other two windings by 180° and in the star or Y connection this puts phase voltages in series which are only 60° apart. So when two equal voltages which are 60° apart are connected in series, their total voltage at any instant will not be double but will be approximately 1.73 (the square root of 3) times the voltage of either one.

To better understand how the sum of polyphase voltages are derived, vector diagrams are sometimes used. The one in Fig. 11-13, for example, shows how polyphase values are obtained. In general, the lines are drawn to scale and at the proper angle to represent the voltages to be added. The line from *B* to *A* represents 120 V of one winding, and the line from *B* to *C* represents 120 V to another winding 120° out of phase with the first. However, as one of the phases is reversed with respect to the other, the line is drawn in the opposite direction from *B* to *C* to represent the voltage 180° displaced or in the reverse direction to that shown by line *BA*. This voltage will then be 60° displaced from that in the other phase, shown by line *BC*.

By completing the parallelogram of forces as shown by the dotted lines, the vectorial sum of the two-phase winding voltages can now be determined by measuring the diagonal line *BE*. If the lengths of the lines *BC* and *BD* are each allowed to represent 120 V by a scale of ⅛ in. for each 10 V, the length of line *BE* measures out to 1.73 times as long as either of the others, so it will represent about 208 V.

The important fact to remember when working with polyphase circuits is that the star or Y connection always gives 1.73 (1.732 to be exact) times the voltage of the delta connection. So in changing from delta to star, multiply the delta line voltage by 1.732, and in changing from star to delta, divide the star line voltage by 1.732, or multiply it by .5774, to obtain the delta line voltage.

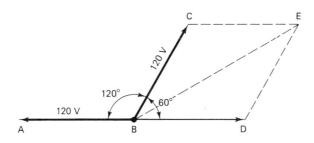

FIG. 11-13   (Courtesy Page Power Co.)

12

Selection of Electric Motors for the Job

For an electric motor to perform efficiently, the proper type of motor and control for the application must be selected. To make such a selection, a knowledge of the characteristics of ac and dc motors and controls is essential. It is also essential to have a knowledge of the general operating characteristics of the load the motor will operate, such as power and speed requirements, special operating conditions, control features, and similar items.

TYPES OF DRIVES

In general, two methods of drive have been traditional: using one motor to drive two or more machines as a group or by having an individual motor for each machine. While the former was the popular method decades ago, the latter method is the most popular today. In the group drive system, the various machines are connected together by shafting and belts as shown in Fig. 12-1, although chains, gears, or other mechanical devices may be used.

Seldom will a group drive system be seen in any modern industrial or commercial application. Practically all systems are now of the individual drive type (Fig. 12-2). The reasons are many, but flexibility of location or arrangement of the machine with individual motor drive makes it pos-

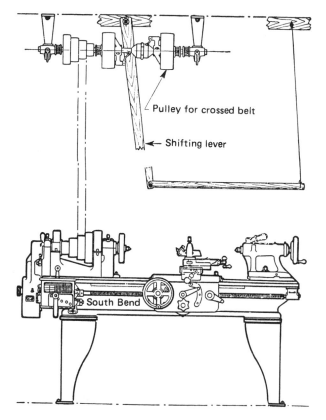

Pulley for crossed belt

← Shifting lever

South Bend

FIG. 12-1 The group drive system is all but obsolete, giving way to chains, gears, and other mechanical devices. (Courtesy Page Power Co.)

sible to place each machine in the best position to suit the flow of material and save handling and trucking. Also, as manufacturing conditions change, each machine – complete with its drive and control as a unit – may be readily moved to some other location. The elimination of overhead belting and shafting greatly improves the lighting on the machines and also facilitates the use of material-handling equipment, such as conveyor systems, small cranes, and similar devices. The most important factor, however, is probably the ease with which machines using individual motors may be started and stopped. Also, with this system, the exact speed required may be obtained and maintained indefinitely, being independent of the load on other machines and of temperatures and atmospheric conditions, which may affect belt transmission.

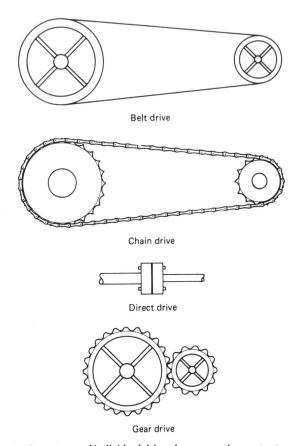

Belt drive

Chain drive

Direct drive

Gear drive

FIG. 12-2   Some types of individual drives for connecting motor to machine.

SELECTING THE MOTOR FOR THE JOB

When selecting motors for a particular application, no set rules will fit each and every application, but there are some important machine and load characteristics that should be known for practically any and all requirements:

1. The speed requirement of the driven shaft, in rpm
2. The range of speed required, if the machine is adjustable
3. The horsepower required at maximum speed or loading
4. If the load is not constant, the cycle duty, including variable items, such as load, time, speed, weights, and other factors
5. If the speed varies, the variation of the torque with the speed

6. Torque: starting, pull-in, pull-out, or maximum – all in percent of full-load values

7. Mechanical connection: belt, chain, coupling, or gear.

DETERMINING TORQUE. It is sometimes desirable to check the manufacturer's data giving the motor characteristics. This is done by installing a temporary motor and taking power readings under various operating conditions.

The starting torque required can be determined by wrapping a rope around the driven pulley and then measuring, with a spring scale, the pull which will start the machine and turn it over. The starting torque, in pound-feet, is the product of the reading on the scale, in pounds, and the radius of the pulley, in feet. For example, assume that it requires a pull of 75 lb on a driven pulley of 6-in. radius to start the motor. The starting torque required is

$$75 \times \tfrac{6}{12} = 3.75 \text{ lb} \cdot \text{ft at 1-ft radius}$$

In most cases, however, the manufacturer's data can be relied upon and will suffice for most applications when selecting a motor for a particular job.

POWER SUPPLY

If the power supply is alternating current, it is necessary to know the frequency, voltage, and number of phases. However, if the voltage is direct current, only the voltage need be known.

The characteristics of the electric service and its limitations must be considered in every instance when selecting motors for a given application.

LOAD FACTOR

The load factor is the ratio of the average load to the maximum load over a certain period. The time may be either the normal number of operating hours per day or may be 24 h. The average load is equal to the kilowatt-hours used in the specified time as measured by watt-hour meters divided by the number of hours. The maximum load is the highest load at any one time as measured by some form of maximum-demand or curve-drawing watt-hour meter.

When comparing industrial loads, the maximum load is taken as the load which would be obtained if all motors were operating continuously at the full-rated load for the same period. This input to the motors is obtained by using the equation

$$\text{full-load input, in kW} = \frac{\frac{\text{total hp rating}}{\text{of motors} \times .746}}{\frac{\text{average efficiency}}{\text{of motors}}} \times \text{hours of operation per day}$$

If the load factor is based on the number of working hours per day, the 24-h load factor may be obtained by multiplying the given load factor by the number of hours and dividing by 24.

TYPES OF MOUNTING AND ENCLOSURES

Two important mechanical characteristics must be considered when selecting motors for a particular application. One is the type of mounting to be used, and the other is the type of enclosure and ventilation.

The types of motor mounting to consider are the horizontal, vertical, flange, gear motor, and others. Common types of enclosures include open drip-proof, totally enclosed fan-cooled, totally enclosed explosion-proof, and separately vented—pipe vent or blower.

OPEN DRIP-PROOF. This design draws outside cooling air into the motor for ventilation. It is primarily for clean, dry areas indoors. Contaminated cooling air will normally reduce the life of the motor (winding and bearing grease). This motor is good for general-purpose use but not ideal where little maintenance can be performed.

TOTALLY ENCLOSED FAN-COOLED. Outside cooling air is directed over the motor by an exterior shaft-mounted fan. It is ideal for dusty atmospheres and many hostile areas. It is primarily used in areas where the motor may not be accessible for maintenance. Cooling is not as efficient as an open-type motor.

TOTALLY ENCLOSED EXPLOSION-PROOF. Ventilation here is the same as the totally enclosed fan-cooled. The motors are specifically designed for installation in hazardous areas as defined by the NE Code. The motor is built to contain and withstand an explosion within its own enclosure. Because highly flammable gases and dust require special design requirements (thermal, mechanical, and thermal protective devices), attention is given to machined fits and mating surfaces to ensure adequate flame paths to contain and extinguish flames or sparks before they reach the outside air.

SEPARATELY VENTED—PIPE OR BLOWER. These motor

types are the least common. Blower cooling is used where high heat-generating duty cycles are found and/or motor size is at a premium. Blower-forced air cooling can remove heat quickly. This design approach can be expensive compared to conventional enclosures and is dependent on heat removal as the key design factor rather than torque limitations. Pipe ventilation is similar except here the motor is generally ducted to a supply of clean, fresh air. It may or may not have a blower. Often the motor may require oversizing here where the duct is long. Maintenance is critical to each of these to ensure that ventilation passages are clear. The application considerations that follow can be applied to the basic enclosures outlined.

HIGH-SPEED BELTING

An ever-increasing number of processes require higher speed for greater output or simply to properly blend special materials being mixed. This, coupled with the fact that higher-capability fiberglass belts are often used, can reduce bearing and shaft life.

High-speed belting usually employs a 3600-rpm motor. Conventional bearings operating at high speed and high load factors begin to approach their design point in limiting speed. Lubrication breakdown is also prevalent.

Drive considerations as a minimum should utilize ductile iron, dynamically balanced sheaves, and matched belts. Transmittal of this information to the motor supplier with the belt drive details (center distance, sheave pitch diameters, etc.) is critical if a marriage of the motor drive is to be successful.

The bearing should be selected to maximize life, and this can mean a standard ABEC-1 commercial-motor-quality standard bearing, an ABEC-3 for truer tolerances on the ball and shaft bore, or a bronze or phenolic retainer for the balls to increase speed capability. It is critical not to neglect the limiting bearing speed recommended by the bearing manufacturers.

Lubrication is generally grease of a No. 2 grade. Oil lubrication may be a consideration as well as oil mist. Grease is a more economical approach and is the first choice.

Shaft deflection imposed by heavy belt tensions or overtensioning can mean oversize shafting will be required as well as increased diametrical clearance between the shaft and bearing housing to prevent rubbing. A common misunderstood fact is that changing to higher-level psi shaft material will not provide a stiffer, lesser deflecting assembly:

$$\text{deflection} = \frac{WI^3}{48EI}$$

where

$$W = \text{load}$$
$$I = \text{length of shaft}$$
$$E = \text{modulus of elasticity} \left(\frac{\text{stress}}{\text{strain}} \right)$$
$$I = \text{moment of inertia}$$

Diameter changes have the greatest impact upon shaft deflection; the modulus of elasticity is the same or near the same for almost all common motor shaft materials and has little effect upon deflection.

Recommendations:
1.   Check bearing life in hours for radial shaft load.
2.   Verify bearing speed capability.
3.   Use balanced, ductile sheaves and matched belts.
4.   Check standard shaft deflection, and change diameter if necessary.
5.   Open shaft bearing-housing clearance.
6.   Be sure belt drive is tensioned properly (glass belts have reduced deflection rates for tensioning).

DUTY CYCLE

Because of special product requirements, motors are subjected to duty cycles – intermittent operation with frequent starting, stopping, and often reversing. The associated heating created by high loads for short periods of time often causes the need for higher insulation classes and/or blower cooling.

An approximate determination of required motor horsepower can be made from a cyclic load curve using the rms (root mean square) method, which is an arithmetic integration of the square of the load curve as follows:

$$\text{rms} = \sqrt{\frac{\text{hp}_1^2 (T_1 - T_0) + \text{hp}_2^2 (T_2 - T_1) + \cdots + \text{hp}_{10}^2 (T_{10} - T_9)}{T_{10} - T_0}}$$

hp is an average hp between T_0 (time$_0$) and T_{10}, etc. After arriving at the final rms hp value, it is necessary to take the peak horsepower value encountered during the cycle and convert it to lb·ft. Then this value can be plotted on the motor speed torque curve to be sure the motor can produce this torque without stalling.

Recommendations:
1. Calculate rms horsepower.
2. Select a motor with one higher insulation class to cover peak loads.

There are numerous cases where reversing is a part of the duty cycle. Reversing capacity is rated in idle reversals that can be performed with no load or connected inertia. Of course, a motor will always see some connected inertia. Since the reversing capacity of a motor is inversely proportional to the total connected inertia, the following applies:

$$\frac{Rx}{Ri} = \frac{WK^2 r}{WK^2 t}$$

where

$$Rx = \text{reversal acceleration with connected inertia}$$

$$Ri = \text{reversal acceleration idle}$$

$$WK^2 r = \text{rotor inertia (lb} \cdot \text{ft}^2)$$

$$WK^2 t = \text{total connected inertia (load and rotor)}$$

Reversal capacity is limited by the amount of heat generated by the reversal itself in a particular frame size. Once the data in the preceding formula are determined, reversals can be calculated. All that remains is to assign values of heat units to the cycle. A heat unit is a segment of the cycle to which a heating increment is assigned.

$$\text{motor acceleration to full speed} = 1 \text{ heat unit}$$

$$\text{dc brake stop} = 1 \text{ heat unit}$$

$$\text{plug stop} = 3 \text{ heat units}$$

$$\text{plug reversal} = 4 \text{ heat units}$$

$$\text{running losses, inertia, inaccuracies, and manufacturing tolerances} = \text{heat units}$$

By selecting the applicable cycle portions listed, total heat units can be easily calculated. Dividing the connected inertia reversal rate Rx by the heat units per cycle yields the loaded cycles per minute.

Recommendations:
1. Low inertia rotors in high reversal rate designs
2. High stator copper content for lower motor temperatures
3. Pinned or keyed rotors for mechanical resistance to reversing

4. Aluminum vent fan keyed to shaft for positive cooling and mechanical resistance to reversing

INERTIA

High-inertia drive motors are those capable of accelerating very large loads from rest to full speed. Typically, these loads are fans or centrifuges.

The primary consideration is that of heat dissipation. During starting, heat is generated inside the motor rotor and stator. The degree and effectiveness of heat dissipation is a direct function of the magnitude of inertia that can be accelerated.

During starting, most heat is generated in the motor rotor and stator due to high inrush current and high rotor slip. Since this heat is directly related to the inertia accelerated, the motor must then be able to absorb the heat until heat transfer takes place, allowing normal ventilation to take over. Temperature is then directly proportional to pounds of active material (copper and steel). If acceleration time is increased by means of reduced voltage starting, then the heat transfer allows the motor to adequately dissipate the heat generated. (Less inrush current reduces the amount of heat that must be dissipated.) Acceleration times will be lengthened considerably – often from 1 to 2 min up to 8 to 10 min.

The important criterion to consider here is that the available torque at reduced voltage must exceed the friction and windage torque of the drive by an ample amount to provide adequate acceleration. Torque becomes the important design factor rather than horsepower as well as start-stop requirements. Star-delta starting as compared to full-voltage starting would yield the following different motor capabilities:

75-hp, 1800-rpm, Enclosed Motor

Motor Inertia	Star-Delta Start	Full-Voltage Start
Capability	6600	2600
Acceleration		
Time	10 min	2 min

Reduced voltage starting will typically reduce stator temperature to as low as 50–60% of the line start value. This is often desirable for process requirements in air-conditioned facilities as well as for substantially extending motor life.

Thermal protection is highly desirable, and two sets of protectors are recommended. The first set of protectors is for starting. They are for the maximum temperature condition to eliminate nuisance tripping for two consecutive starts of the high inertia with the motor at operating temperature. The second set of protectors is for the running mode, and they are set for normal rated running temperature. The motor insulation is rated for the maximum rated temperature condition so as to prolong life.

Recommendations:

1. Define inertia and friction and windage.

2. Use reduced voltage starting if inertia values are high (higher than across the line start).

3. Verify motor accelerating torque.

4. Define starting cycles.

5. Define duty cycle.

6. Check starting time.

7. Identify belting (most are belted or geared – direct drive is seldom used) so it can be checked.

8. Use thermal protection for starting and running condition.

SHOCK LOAD

Often a motor will be installed in an area subjected to heavy shock loads. They can range from punch presses to drop forges to vibrating screens and conveyors. It is also common in power plants to design motors for loads imposed by earthquake.

Obviously, magnitude and frequency weigh heavily on the ultimate design considerations. Usually, the stator winding will be most vulnerable to failure by shock load and failure due to grounding or turn-to-turn short circuiting.

The basic motor frame selection should be for high strength – cast iron housing, bearing brackets, and fan guards.

Recommendations:	
Design Condition	*Solution*
1. Occasional mild vibration	1. Additional varnish bakes to add mass and rigidity to stator end coils
2. Frequent moderate vibration	1. Additional varnish bakes; see previous solution
	2. Bonded insulating tape wrapped around end coils or laced end coils

	3. Cast iron conduit box
	4. Additional varnish bakes on all hardware
	5. Vent fan keyed to shaft
	6. Rotor keyed to shaft – not shrunk
3. Frequent heavy vibration	Same as items 1-6, above, plus:
	7. Ductile iron castings
	8. Laced end coils plus overcoat of heavy epoxy resin
	9. Clamped or packed leads in housing lead channel

It is often necessary to also consider oversize shafting, special bearings, and heavier-grade grease for the most severe cases. Special bearings usually result in the use of a special heavy-duty ABEC-3 fit roller bearing or spherical roller bearing. Greases frequently used are those with high film strength.

RADIATION

Radiation requirements are most generally associated with nuclear power plants. However, some test labs use radioactive elements as well and therefore have need of motors capable of withstanding radiation dosage.

It is of prime importance that insulating components be of inorganic origin – wire varnish, top wedge, slot liners, phase insulation, and leads. Most frequently silicons, glass, and mica are utilized. The other major consideration is bearing grease. Here again, a silicone base is most often used.

One item that cannot be neglected is the insulating or core plate often used on the lamination steel. Again, a silicone product is selected.

Recommendations:

1. Define radiation levels and submit to the motor supplier. Quantify levels, generally in rads of 1×10^x.

2. State required design life necessary in years, if applicable.

3. Work with the motor manufacturer to select a suitable lubricant that is readily available for normal maintenance.

TEMPERATURE

Most users are knowledgeable of effects of higher-range temperatures on motor life—specifically the stator insulation. A rule of thumb is that motor insulation life is halved for every additional 10°C of operating temperature. However, with today's higher-grade insulation materials, motors can be built with very acceptable life while operating at higher temperature values.

The three most common conditions causing higher motor temperature are the following:

1. High ambient
2. High altitude
3. Service factor (multiplier applied to normal nameplate hp ratings)

Factors such as overload and single phasing are not considered here since they are not a design consideration to be covered by the continuous operation mode.

Ambient (°C)	Altitude (ft)	Insulation Class	Temp. Rise (°C)
0–40	0–3300	Class B	80
40–65	0–3300	Class F	105
65–85	0–3300	Class H	125
0–40	3300–9900	Class F	105
0–40	9900–12,000	Class H	125

Even higher-temperature operation can be designed for if necessary. Considerations then are given to very high-temperature insulation (200°C total temperature), very special grease, loose motor fitups to permit thermal expansion, and heat-stabilized bearings. The motor is usually oversized to allow for lower temperature rise and permit adequate life. Frequent maintenance of bearings is mandatory.

CORROSION

Corrosion is the most frequently occurring condition—the severity and thus the recommendation depend on the actual motor location. Most motors in highly corrosive atmospheres whether acids, bases, or salts will have markedly increased life by utilization of a cast iron exterior. Additions to this may be as simple as exterior epoxy paint or as special as stainless steel housing shrouds.

Frequently, motors will operate where condensation within the motor occurs due to corrosive contaminant vapors condensing or simple moisture from humidity condensing. The best and most economical means of correcting for these common conditions are the following:

1. Drains to relieve excess moisture

2. Epoxy to protect motor windings from caustic vapor attack

3. Space heaters to keep the motor interior temperature above the dew point

Exterior paints are available for virtually any special application. The very exotic types such as vinyls and coal tar epoxy, to name two, are a small percentage for even the most demanding situation. Most applications can be met with the standard paint used by most motor manufacturers with the remaining portion adequately handled by a standard epoxy-type overcoat.

The small percentage of exterior paints mentioned usually requires sandblasting of parts prior to priming and overcoats of paint in several stages at prescribed intervals for proper bonding and adhesion. Mixing of the paint is often critical for proper application in specific amounts. When these types are utilized, it is mandatory that other equipment considerations such as the motor interior (winding, grease, rotor, shafting, seals, etc.) be examined in detail to verify their adequacy, the point being that the exterior appearance is of little value if the motor experiences premature internal failure.

Seal selection is another prime factor to ensure that contaminants are kept outside the motor. This refers not only to the shaft entrance but bearing housing fit and conduit box fitups as well.

For severe conditions, the following is recommended:

Motor Area	Modification
Housing to bearing bracket fits	'O' Ring seals
Shaft	Stainless material spring-loaded seals where moisture present, rotating seals (rubber) where dictated by environment
Conduit box	Gaskets
Housing – lead channel entrance to conduit box	Packing material

Ventilating fans are critical where enclosed fan-cooled motors are used. Three basic types are common: polypropylene (plastic), aluminum, and brass. Where they are unacceptable, highly resistant epoxy coatings can be applied to provide very acceptable life to this critical part.

Conductive dusts are another frequently encountered environment. Attention is focused here on the stator winding end turn to protect the winding from being coated or covered with the material that can short the stator circuit. An overcoat of polyurethane dispersed particles to a 7-mil buildup is applied. This provides excellent resistance to abrasion as well.

ACKNOWLEDGMENTS

I am indebted to Mr. P. D. Preuninger, Special Products Manager, Louis Allis Division, Litton Industrial Products, Inc, for supplying much reference material used in this chapter. If you have a particular application that is not covered herein, it is suggested that you contact Louis Allis Division at 427 East Stewart St., Milwaukee, WI 53201 for additional information.

13

Motor Installation

The best motors on the market will not operate properly if they are installed incorrectly. Therefore, all personnel involved with the installation of electric motors should thoroughly understand the proper procedures for installing the various types of motors that will be used. Furthermore, proper maintenance of each motor is essential to keep it functioning properly once it is installed.

When an electric motor is received from the manufacturer or supplier, always refer to the manufacturer's instructions and follow them to the letter. Failure to do so could result in serious injury or fatality. In general, disconnect all power before servicing. Install and ground according to the NE Code and all local codes. Consult qualified personnel with any questions or services required.

UNCRATING

Once the motor has been carefully uncrated, check to see if any damage has occurred during handling. Be sure that the motor shaft and armature turn freely. This time is also a good time to check to determine if the motor has been exposed to dirt, grease, grit, or excessive moisture in either shipment or storage before installation. Motors in storage should have shafts turned over once each month to redistribute grease in the bearings.

Never start a motor which has been wet without having it thoroughly dried.

The measure of insulation resistance is a good dampness test. Clean the motor of any dirt or grit.

SAFETY

Eyebolts or lifting lugs on motors are intended only for lifting the motor and factory motor-mounted standard accessories. These lifting provisions should never be used when lifting or handling the motor and other equipment such as pumps, gear boxes, fans, or other driven equipment as a single unit.

The eyebolt lifting capacity rating is based on a lifting alignment coincident with the eyebolt centerline. The eyebolt capacity reduces as deviation from this alignment increases.

All motors should be installed, protected, and fused in accordance with the latest issue of the NE Code, NEMA Standard Publication No. MG-2, and any and all local requirements.

Frames and accessories of motors should be grounded in accordance with the NE Code, Article 430. For general information on grounding, refer to the NE Code, Article 250.

Rotating parts such as pulleys, couplings, external fans, and unusual shaft extensions should be permanently guarded against accidental contact with clothing or body extremities.

THERMAL PROTECTOR INFORMATION

A space on a nameplate will be stamped or not be stamped to indicate the following:

1. The motor is thermally protected.
2. The motor is not thermally protected.
3. The motor has an overheat protective device.

For examples, refer to the following paragraphs:

1. Motors equipped with built-in thermal protection have *thermally protected* stamped on the nameplate. Thermal protectors open the motor circuit electrically when the motor overheats or is overloaded. The protector cannot be reset until the motor cools. If the protector is automatic, it will reset itself. If the protector is manual, press the red button to reset.

2.   Motors without thermal protection have nothing stamped on the nameplate about thermal protection.

3.   For motors that are provided with an overheat protective device that does not open the motor circuit directly the nameplate will be stamped *with overheat protective device*.

   a.   Motors with this type of overheat protective device have protector leads brought out in the motor conduit box marked P_1 and P_2. These leads are intended for connection in series with the stop button of a three-wire pilot circuit of a magnetic controller to a motor, as in Fig. 13-1.

   b.   The load controlled by the preceding overheat protective device cannot exceed the values shown in the following table:

ac Volts	Volt-amp Ratings	ac Volts	Volt-amp Ratings
120	360	208	360
240	360	480	360
600	360		

LOCATION

In selecting a location for the unit, first consideration should be given to ventilation. It should be far enough from walls or other objects to permit a free passage of air.

   The motor should never be placed in a room with a hazardous pro-

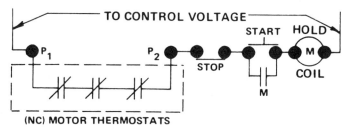

AC VOLTS	VOLT–AMP RATINGS	AC VOLTS	VOLT–AMP RATINGS
120	360	208	360
240	360	480	360
600	360		

FIG. 13-1   Motor protector leads are intended for connection in series with the stop button of the magnetic controller. (Courtesy Marathon Electric.)

cess or where flammable gasses or combustible material may be present unless it is specifically designed for this type of service.

1. Drip-proof motors are intended for use where the atmosphere is relatively clean, dry, and noncorrosive. If the atmosphere is not like the preceding, then request approval of the motor for the use intended.

2. Totally enclosed motors may be installed where dirt, moisture, and corrosion are present or in outdoor locations.

3. Explosion-proof motors are built for use in hazardous locations as indicated by the Underwriters' label on the motor. Consult your local governmental inspection agency for guidance.

The ambient temperature of the air surrounding the motor should not exceed 40°C or 104°F unless the motor has been especially designed for high-ambient-temperature applications. The free flow of air around the motor should not be obstructed.

After a location has been decided upon, the mounting follows. For floor mounting, motors should be provided with a firm, rigid foundation, with the plane of four mounting stud pads flat within .010 in. for a 56 to 210 frame and .015 in. for a 250 to 680 frame. This may be accomplished by shims under the motor feet.

MOTOR DRIVES

Before connecting the motor to the load by a belt drive or direct coupling, verify manually that the rotor turns freely and does not rub.

V-BELT DRIVE.

1. Align the sheaves carefully to avoid axial thrust on the motor bearing. The drive sheave on the motor should be centered on the shaft extension.

2. Adjust the tension just enough to prevent excessive bow of the slack side.

3. If possible, make the lower side of the belt the driving side.

4. The pulley ratio should not exceed 8 to 1 or as approved by the manufacturer for a specific application.

5. Sheaves should be in accordance with NEMA Spec. MG-1.

DIRECT-CONNECTED DRIVE.   Flexible or solid shaft couplings must be properly aligned for satisfactory operation. On flexible couplings, the clearance between the ends of the shafts should be in accordance with the coupling manufacturer's recommendations or NEMA standards for end play and limited travel in coupling.

Angular misalignment and run-out between direct-connected shafts will cause increased bearing loads and vibration even when the connection is made by means of a flexible coupling.

To check for angular misalignment, clamp the dial indicator to one coupling hub and place the finger or button of the indicator against the finished face of the other hub as shown in Fig. 13-2(a). Set the dial at zero.

Rotate one shaft, keeping the indicator button at the reference mark on the coupling hub, and note the reading on the indicator dial at each revolution.

Angular misalignment of the shafts must not exceed a total indicator reading of .002 in. for each inch of diameter of the coupling hub.

After the shafts have been checked for angular misalignment and are parallel within the limits specified in the preceding paragraph, check the shaft for run-out to assure concentricity of the shafts. Clamp the indicator button on the machined diameter of the other hub as shown in Fig. 13-2(b). Set the dial at zero.

Rotate one shaft, keeping the indicator button at the reference mark on the hub, and note the reading on the indicator dial at each revolution.

Total run-out between the hubs should not exceed .002 in.

Rotating parts such as couplings, external fans, and unusual shaft extensions should be permanently guarded against accidental contact with clothing or body extremities.

ELECTRICAL CONNECTIONS

All wiring, fusing, and grounding must comply with the NE Code and local requirements.

To determine proper wiring, rotations and voltage connections, refer to the information and diagram on the nameplate, separate connection plate, or decal. If the plate or decal has been removed, make inquiries of the manufacturer.

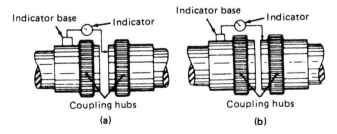

FIG. 13-2 A dial indicator is used to check for angular misalignment in direct-connected drives. (Courtesy Marathon Electric.)

Use the proper size of line current protection and motor controls as required by the NE Code and local codes. Recommended use is 125% of full-load amps as shown on the nameplate for motors with 40°C ambient and a service factor over 1.0. Recommended use is 115% of full-load amps as shown on the nameplate for all other motors. Do not use protection with capacities larger than recommended. All three-phase motors must have all three phases protected.

CHANGING ROTATION

Before a motor may be used as a power source for equipment, the rotation sequence of both the motor and the driven unit must be determined. Rotation may be checked by jogging or bumping by applying power to the motor leads for a very short period of time – enough to just get the motor shaft to rotate a slight amount to enable observation of the shaft rotating direction.

The rotation may be changed on three-phase motors by interchanging any two of the line leads with the motor lead connections.

PART WINDING STARTING

On those motors used for part winding starting, the elapsed time on the first step should be kept as short as possible and should not exceed 5 s. It is recommended that this time be limited to 2 s.

OPERATION

BEFORE INITIAL STARTING.   The following should be checked before a motor is first started:

1. If a motor has become damp in shipment or in storage, it is advisable to measure the insulation resistance of the stator winding. This value should be approximately

$$\text{megohms} = \frac{\text{rated voltage}}{1000} \quad \text{(but no less than .5 M}\Omega\text{)}$$

2. If insulation resistance is low, dry out the moisture in one of the following ways:

   a. Bake in an oven at a temperature of not more than 90°C (194°F) until the insulation resistance is practically constant.

   b. Enclose the motor with canvas or a similar covering, leaving a hole at the top for moisture to escape, and insert the heating units or lamps.

c. Pass a current at low voltage (rotor-locked) through the stator winding. Increase the current gradually until the winding temperature, measured with a thermometer, reaches 90°C (194°F). Do not exceed this temperature.

3. See that the voltage and frequency stamped on the motor and the control nameplates correspond with that of the power line.

4. Check all connections to the motor and control with the wiring diagram.

5. Be sure the rotor turns freely and does not rub when disconnected from the load. Any foreign matter in the air gap should be removed.

6. Leave the motor disconnected from the load for the initial start; it is desirable to operate the motor without load for about 1 h to test for any localized heating in the bearings and windings. Check for proper rotation.

COLLECTOR RINGS (WOUND ROTOR MOTORS ONLY).   The collector rings are sometimes slushed at the factory to protect them while in stock and during shipment. Before putting the motor into service, the slushing should be removed with carbon tetrachloride or some other cleaning fluid that will not attack insulation, the rings should be polished with fine sandpaper, and the brushes should be set down on the collector surface. Keep the rings clean and maintain their polished surfaces. Ordinarily, the rings will require only occasional wiping with a piece of canvas or nonlinting cloth. Do not let dust or dirt accumulate between the collector rings.

BRUSHES (WOUND ROTOR MOTORS ONLY).   See that the brushes move freely in the holders and at the same time make firm, even contact with the collector rings. The pressure should be between 2 and 3 psi of brush surface.

When installing new brushes, fit them carefully to the collector rings. Be sure that the copper pigtail conductors are securely fastened to, and make good contact with, the brush holders.

ALLOWABLE VOLTAGE AND FREQUENCY RANGE.   If the voltage and frequency are within the following range, motors will operate but with somewhat different characteristics than obtained with correct nameplate values:

1. *Voltage:* Within 10% above or below the value stamped on the nameplate

2. *Frequency:* Within 5% above or below the value stamped on the nameplate

3. *Voltage and frequency together:* Within 10% (providing the preceding frequency is less than 5%) above or below values stamped on the nameplate.

It is of absolute importance to keep both the interior and exterior of the motor free from dirt, water, oil, and grease. Motors operating in dirty places should be periodically disassembled and thoroughly cleaned.

If the motor is totally enclosed, fan-cooled, or nonventilated and is equipped with automatic drain plugs, the plugs should be free of oil, grease, paint, grit, and dirt so they do not clog up.

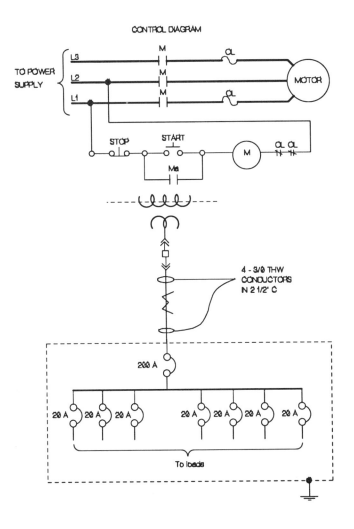

CONTROL DIAGRAM

OVERCURRENT PROTECTION FOR MOTORS

Every fuse has a specific ampere rating. In selecting the ampacity of a fuse, consideration must be given to the type of load and code requirements. The ampere rating of a fuse should normally not exceed the current carrying capacity of the circuit. For instance, if a conductor is rated to carry 20 A, a 20-A fuse is the largest that should be used in the conductor circuit. However, there are some specific circumstances when the ampere rating is permitted to be greater than the current carrying capacity of the circuit. A typical example is the motor circuit; dual-element fuses generally are permitted to be sized up to 175% and non-time-dely fuses up to 300% of the motor full-load current in amperes. Generally, the ampere rating of a fuse and switch combination should be selected at 125% of the load current. There are exceptions, such as when the fuse-switch combination is approved for continuous operation at 100% of its rating.

Guide for Sizing Fuses

General guidelines for sizing fuses are given here for most circuits that will be encountered on conventional electrical systems. Some specific applications may warrant other fuse sizing; in these cases, the load characteristics and appropriate NE Code sections should be considered. The selections shown here are not, in all cases, the maximum or minimum ampere ratings permitted by the NE Code. Demand factors as permitted per the NE Code are not included here.

Dual-Element Time-Delay Fuses

Feeder Circuit With All Motor Loads. Size the fuse at 150% of the full load current of the largest motor plus the full-load current of all motors.

Feeder Circuit With Mixed Loads. Size fuse at sum of 150% of the full-load current of the largest motor plus 100% of the full-load current of all other motors plus 125% of the continuous, non-motor load plus 100% of the non-continuous, nonmotor load.

Motor Branch Circuit With Overload Relays. Where overload relays are sized for motor running overload protection, the following provide backup, ground fault, and short-circuit protection: Motor 1.15 service factor or 40 degrees C. rise; size fuse at 125% of motor full-load current or next higher standard size.

B. Motor less than 1.15 service factor or over 40 degrees C. rise; size the fuse at 115% of the motor full-load current or the next higher standard fuse size.

Motor Branch Circuit With Fuse Protection Only. Where the fuse is the only motor protection, the following fuses provide motor running overload protection and short-circuit protection.

Motor 1.15 service factor or 40 degrees C. rise; size the fuse at 100% to 125% of the motor full load current.

Motor less than 1.15 service factor or over 40 degrees C. rise: size fuse at 100% to 115% of motor full load current.

Large Motor Branch Circuit. Fuse larger than 600 amps. For large motors, size KRP-C HI-CAP time-delay fuse at 150% to 225% of the motor full load current, depending on the starting method; that is, part-winding starting, reduced voltage starting, etc.

Non-Time-Delay Fuses

Feeder Circuit With All Motor Loads. Size the fuse at 300% of the full-load current of the largest motor plus the full-load current of all other motors.

Feeder Circuit with Mixed Loads. Size fuse at sum of 300% of full load current of largest motor plus 100% of full-load current of all other motors plus 125% of the continuous, non-motor load plus 100% of non-continuous, non-motor load.

Branch Circuit With No Motor Load. The fuse size must be at least 125% of the continuous load plus 100% of the non-continuous load. Do not size larger than the ampacity of conductor.

Motor Branch Circuit With Overload Relays. Size the fuse as close to but not exceeding 300% of the motor running full-load current. Provides ground fault and short-circuit protection only.

Motor Branch Circuit With Fuse Protection Only. Non-time-delay fuses cannot be sized close enough to provide motor running overload protection. If sized for motor overload protection, non-time-delay fuses would open due to motor starting current.

When sizing fuses for a given application, a schematic drawing of the system will help tremendously. Such drawings do not have to be detailed, just a single-line schematic, such as the one shown in Fig. 13-3 will suffice.

COORDINATION

"Coordination" is the name given to the time-current relationship among a number of overcurrent devices connected in series, such as fuses in a main feeder, subfeeder, and branch circuits. Safety is the prime consideration in the operation of fuses; however, coordination of the characteristics of fuses has become a very important factor in the large and complex electrical system of the present time. Every fuse should be properly rated for continuous current and overloads and for the maximum short-circuit current the electrical system could feed into a fault on the load side of the fuse—but this is not enough. It might still be possible for a fault on a feeder to open the main service fuse before the feeder fuse opens. Or, a branch-circuit fault might open the feeder fuse before the branch-circuit device opens. Such applications are said to be uncoordinated, or nonselective—the fuse closest to the fault is not faster operating than one farther from the fault. When a fault on a feeder opens the main service fuse instead of the feeder fuse, all of the electrical system is taken out of service instead of just the one faulted feeder. Effective coordination minimizes the extent of electrical outage when a fault occurs. It therefore minimizes loss of production, interruption of critical continuous processes, loss of vital facilities, and possible panic.

Selective coordination is the selection of overcurrent devices with time/current characteristics that assure clearing of a fault or short circuit by the device nearest the fault on the line side of the fault. A fault on a branch circuit is cleared by the branch-circuit device. The sub-feeder, feeder, and main service overcurrent devices will not operate. Or, a fault on a feeder is opened by the feeder fuse without opening any other fuse on the supply side of the feeder. With selective coordination, only the faulted part of the system is taken out of service, which represents the condition of minimum outage.

With proper selective coordination, every device is rated for the maximum fault current it might be called upon to open. Coordination is achieved by studying the curve of current vs. time required for operation of each device. Selection is then made so that the device nearest any load is faster operating than all devices closer to the supply, and each device going back to the service entrance is faster operating than all devices closer to the supply. The main service fuses must have the longest opening time for any branch or feeder fault.

Example: Determine the conductor size, the motor overload protection, the branch-circuit short-circuit and ground-fault protection, and the feeder protection for one 25-horsepower squirrel-cage induction motor (full voltage starting, nameplate current 31.6 amperes, service factor 1.15, Code letter F), and two-horsepower wound-rotor induction motors (nameplate primary current 36.4 amperes, nameplate secondary current 65 amperes 40°C rise), on a 460-volt, 3-phase, 60-Hertz electrical supply.

The full-load current value used to determine the ampacity of conductors for the 25-horsepower motor is 34 amperes [Section 430-6(a) and Table 430-150]. A full-load current of 34 amperes x 1.25 = 42.5 amperes (Section 430-22). The full-load current value used to determine the ampacity of primary conductors for each 30-horsepower motor is 40 amperes [Section 430-6(a) and Table 430-150]. A full-load primary current of 40 amperes x 1.25 = 50 amperes (Section 430-22). A full-load secondary current of 65 amperes x 1.25 = 81.25 amperes [Section 430-23(a)].

The feeder ampacity will be 125 percent of 40 plus 40 plus 34, or 124 amperes (Section 430.24).

Overload. —Where protected by a separate overload device, the 25-horsepower motor with nameplate current of 31.6 amperes, must have overload protection of not over 39.5 amperes [Section 430-6(a) and 43032(a)(1)]. Where protected by a separate overload device, the 30-horsepower motor, with nameplate current of 36.4 amperes, must have overload protection of not over 45.5 amperes

[Section 430-6(a)(1)]. If the overload protection is not sufficient to start the motor or to carry the load, it may be increased according to Section 430-34. For a motor marked "thermally protected," overload protection is provided by the thermal protector [see Sections 430-7(a)(12) and 43032(a)(2)].

14

Maintenance
of Electric Motors

The care given to an electric motor while it is in operation directly affects the life and usefulness of the motor. A motor that receives good maintenance practices will outlast a poorly treated motor many times over. Actually, if a motor is initially installed correctly and it has been properly selected for the job, very little maintenance is necessary – provided it does receive a little care at regular intervals.

CLEANLINESS

The frequency of cleaning motors will depend on the type of environment in which it is used. In general, keep both the interior and exterior of the motor free from dirt, water, oil, and grease. Motors operating in dirty areas should be periodically disassembled and thoroughly cleaned.

If the motor is totally enclosed – fan-cooled or nonventilated – and is equipped with automatic drain plugs, they should be free of oil, grease, paint, grit, and dirt so they don't clog up.

LUBRICATION

Most motors are properly lubricated at the time of manufacture, and it is not necessary to lubricate them at the time of installation. However, if a motor has been in storage for a period of 6 months or longer, it should be relubricated before starting.

To lubricate conventional motors:

1. Stop the motor.
2. Wipe clean all grease fittings (filler and drain).
3. Remove the filler and drain plugs, *A* and *B* in Fig. 14-1.
4. Free the drain hole of any hard grease (use a piece of wire if necessary).
5. Add grease* using a low-pressure grease gun.
6. Start the motor and let it run for approximately 30 min.
7. Stop the motor, wipe off any drained grease, and replace the filler and drain plugs.
8. The motor is ready for operation.

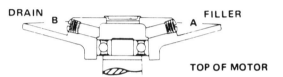

FIG. 14-1   Location of filler and drain plugs in end-bell bearing hub of standard induction motor. (Courtesy Marathon Electric.)

Every 4 years (every year in the case of severe duty) motors with open bearings should be thoroughly cleaned, washed, and repacked with grease. Figure 14-2 shows the relubrication period. Standard conditions reflected in the figure mean 8-h/day operation, normal or light loading, clean at 100°F maximum ambient. Severe conditions are 24-h/day operation, or shock loadings, vibration, or in dirt or dust, clean at 100–150°F ambient, and extreme conditions are defined as heavy shock or vibration, dirt or dust, clean at 100–150°F ambient.

The quantity of grease is important. The grease cavity should be filled one third to one half full. Always remember that too much grease is as detrimental as insufficient grease. Figure 14-3 shows the amount of grease required, and Fig. 14-4 gives the recommended greases.

PRACTICAL MAINTENANCE TECHNIQUES

The key to long, trouble-free motor life—once the motor has been sized and installed properly—is proper maintenance. Maintaining a motor in good operating condition requires periodic inspection to determine if any faults exist and then promptly correcting these faults. The frequency and

---

*The amount of grease added is very important. Only enough grease should be added to replace the grease used by the bearing. Too much grease can be as harmful as insufficient grease.

Frame Size @ 900, 1200 & Var. Speed	Relub. Period @ Std. Conditions (8 hr./day, normal to light loading 100° F max. amb.)	Severe Conditions	Extreme Conditions
140−180	4.5 Years	18 Months	9 Months
210−280	4 Years	16 Months	8 Months
320−400	3.5 Years	14 Months	7 Months
440−508	3.0 Years	12 Months	6 Months
510	2.5 Years	11½ Months	6 Months
Frame Size @ 1800 RPM	Std. Conditions	Severe Conditions	Extreme Conditions
140−180	3 Years	1 Year	6 Months
210−280	2.5 Years	10½ Months	5½ Months
320−400	2.0 Years	9 Months	4½ Months
440−508	1.5 Years	8 Months	4 Months
510	1 Years	6 Months	3½ Months
All Motors over 1800 RPM	6 Months	3 Months	3 Months

FIG. 14-2 Typical relubrication periods for various sizes and types of motors. (Courtesy Marathon Electric.)

thoroughness of these inspections depend on such factors as the following:

- Number of hours and days the motor operates
- Importance of the motor in the production scheme
- Nature of service
- Environmental conditions

Each week every motor in operation should be inspected to see if the windings are exposed to any dripping water, acid, or alcoholic fumes as well as excessive dust, chips, or lint on or about the motor. Make certain that objects that will cause problems with the motor's ventilation system are not placed too near the motor and do not come into direct contact with the motor's moving parts.

In sleeve-bearing motors, check the oil level frequently (at least once a week) and fill the oil cups to the specified line with the recommended lubricant. If the journal diameter is less than 2 in., always stop the motor before checking the oil level. For special lubricating systems, such as forced, flood and disc, and wool-packed lubrication, follow the manufacturer's recommendations. Oil should be added to the bearing housing only when the motor is stopped, and then a check should be made to ensure that no oil creeps along the shaft toward the windings where it may harm the insulation.

BEARING NUMBER	AMOUNT (IN.3)	APPROX. EQUIV. TEASPOONS	BEARING NUMBER	AMOUNT (IN.3)	APPROX. EQUIV. TEASPOONS
203	.15	.5 Tsp.	222	3.0	10.0 Tsp.
205	.27	.9 Tsp.	307	.53	1.8 Tsp.
206	.34	1.1 Tsp.	308	.66	2.2 Tsp.
207	.43	1.4 Tsp.	309	.81	2.7 Tsp.
208	.52	1.7 Tsp.	310	.97	3.2 Tsp.
209	.61	2.0 Tsp.	311	1.14	3.8 Tsp.
210	.72	2.4 Tsp.	312	1.33	4.4 Tsp.
212	.95	3.1 Tsp.	313	1.54	5.1 Tsp.
213	1.07	3.6 Tsp.	314	1.76	5.9 Tsp.
216	1.49	4.9 Tsp.	316	2.24	7.4 Tsp.
219	2.8	7.2 Tsp.	318	2.78	9.2 Tsp.

FIG. 14-3 Typical amount of grease required when regreasing an electric motor. (Courtesy Marathon Electric.)

INSULATION CLASS SHOWN ON NAMEPLATE	GREASE DESIGNATION	GREASE SUPPLIER
B or F	Chevron SRI-2	Standard Oil of California or equivalent

FIG. 14-4 Typical recommended greases for motor lubrication. (Courtesy Marathon Electric.)

Always be alert to any unusual noise which may be caused by metal-to-metal contact (bad bearings, etc.), and also learn to detect any abnormal odor which might indicate scorching insulation varnish.

Feel the bearing housing each week for evidence of vibration and listen for any unusual noise. A standard screwdriver with the blade on the bearing housing and the handle clasped with the hand while the ear is positioned so as to rest on the cupped hand will magnify the noise. Also inspect the bearing housings for the possibility of creeping grease on the inside of the motor which might harm the insulation.

Commutators and brushes should be checked for sparking and should be observed through several cycles if the motor is on cycle duty. A stable copper oxide carbon film – as distinguished from a pure copper surface – on the commutator is an essential requirement for good commutation. Such a film, however, may vary in color from copper to straw or from chocolate brown to black. The commutator should be clean and smooth and have a high polish to prevent problems. All brushes should be checked for wear, and connections should be checked for looseness. The commutator surface may be cleaned by using a piece of dry canvas or other hard, nonlinting material which is wound around and securely fastened to a wooden stick and held against the rotating commutator.

The air gap on sleeve-bearing motors should be checked frequently, especially if the motor has recently been rewound or otherwise repaired. After new bearings have been installed, for example, make sure that the average reading is within 10%, provided the reading should be less than .020 in. Check the air passages through punchings and make sure they are free of all foreign matter.

Compressed air may be used to blow motor windings clean, provided too much pressure is not used. Industrial-type vacuum cleaners have also been used with success. Before performing either of these cleaning operations, however, make certain that the motor is disconnected from the line. Then the windings may be wiped off with a dry cloth. In doing so, check for moisture, and see if any water has accumulated in the bottom of the motor frame. Check also to see if any oil or grease has worked its way up to the rotor or armature windings. If so, clean with AWA 1,1,1 or a similar cleaning solution.

When performing any of the preceding maintenance operations, check other motor parts and accessories such as the belt, gears, flexible couplings, chain, and sprockets for excessive wear or improper location. Also check the starter and that the motor comes up to proper speed each time it is started.

Once every month or so maintenance personnel should check the shunt, series, and commutating field windings for tightness. Do this by trying to move the field spools on the poles, as drying out may have caused some play. If this condition exists, the motor(s) should be serviced immediately. Also check the motor cable connections for rightness and tighten if necessary.

At the same time the preceding checks are made, also check the brushes in their holders for fit and free play. The brush spring pressure should also be checked. Tighten the brush studs in the holders to take up slack from the drying out of washers, making sure that studs are not displaced, particularly on dc motors. All worn or damaged brushes should be replaced at this time. Look for chipped toes or heels and for heat cracks during the inspection.

Each month examine the commutator surface for high bars and high mica or evidence of scratches or roughness. See that the risers are clean and have not been damaged in any way.

Where motors are subjected to hard use, all ball or roller bearing motors should be serviced by purging out the old grease through the drain hold and applying new grease once each month or more frequently if circumstances dictate the need. After each grease change, check to make sure grease or oil is not leaking out of the bearing housing. If so, correct this condition before starting the motor, or the insulation may be damaged.

Check the sleeve bearings for wear about six to eight times each year. Clean out oil wells if there is evidence of dirt or sludge. Flush with lighter oil before refilling.

For motors with enclosed gears, open the drain plug and check the oil flow for the presence of metal scale, sand, grit, or water. If the condition of the oil is bad, drain, flush, and refill as recommended by the manufacturer of the motor. Rock the rotor to see if slack or backlash is increasing.

Loads being driven by motors have a tendency to change from time to time due to wear on the machine or the product being processed through the machine. Therefore, all loads should be checked from time to time for a changed condition, bad adjustment, and poor handling or control.

During the monthly inspection, note if belt-tightening adjustments are all used up. If they are, the belts may be shortened. Also see if the belts run steadily and close to the inside edge of the pulley. On chain-driven machines, check the chain for evidence of wear and stretch and clean the chain thoroughly. Check the chain-lubricating system and note the incline of the slanting base to make sure it does not cause oil rings to rub on the housing.

Once or twice each year, all motors in operation should be given a thorough inspection consisting of the following:

Windings: Check the insulation resistance by using the instruments and techniques described in Chapter 17. The windings should also be given a visual inspection; look for dry cracks and other evidence of a need for coating insulating material. Clean all surfaces thoroughly, especially ventilating passages. Also examine the frame to see if any mold is present or if water is standing in the bottom. Either will suggest dampness and may require that the windings be dried out, varnished, and baked.

Air gap and bearings: Check the air gap to make sure that average readings are within 10% provided the readings should be less than .020 in. All bearings, ball, roller, and sleeve, should be thoroughly checked and defective ones replaced. Waste-packed and wick-oiled bearings should have waste or wicks renewed if they have become glazed or filled with metal, grit, or dirt, making sure that new waste bears well against the shaft.

Squirrel cage rotors: Check for broken parts or loose bars as well as evidence of local heating. If the fan blades are not cast in place, check for loose blades. Also look for marks on the rotor surface which indicate the presence of foreign matter in the air gap or else that the motor has a worn bearing or bearings.

Wound rotors: Wound rotors should be cleaned thoroughly, especially around collector rings, washers, and connections. Tighten all connections. If rings appear to be rough, spotted, or eccentric, they should be refinished by qualified personnel. Make certain that all top sticks or wedges are tight; tighten those which are not.

Armatures: Clean all armature air passages thoroughly. In doing so, look for oil or grease creeping along the shaft back to the bearing. Check the commutator for its surface condition, high bars, high mica, or eccentricity. If necessary, turn down the commutator to secure a smooth fresh surface. This operation is performed in a lathe of suitable size as shown in Fig. 14-5.

For armatures with drilled center holes, put a lathe dog on the shaft opposite the commutator and tighten it. If it is necessary to put the lathe dog on a bearing surface, put a piece of thin copper around the shaft so it

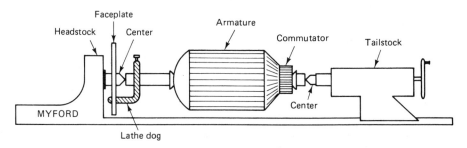

FIG. 14-5   Motor armature secured in lathe between centers.

will not be injured. Put a faceplate on the spindle end of the lathe along with centers in both the headstock and tailstock. Apply some white lead or oil on the tailstock center and then place the commutator between centers and tighten the tailstock, but not so tight as to spread the end of the shaft.

Use a sharp-pointed lathe cutting tool in the tool holder to turn the commutator down, running lathe at medium speed, say, around 700 rpm. Finish the job with a fine file and abrasive paper.

Armatures without drilled center holes will necessitate the use of chucks. The armature shaft opposite the commutator should be chucked in a three-jaw universal chuck in the headstock. Chuck a bearing of proper size in a drill-type chuck and place it in the tailstock of the lathe. Oil this bearing and then place the commutator end shaft into it. See Fig. 14-6. The commutator is then turned in the same manner as described previously.

To summarize, when turning down an armature, always use a pointed tool with a sharp and smooth edge to obtain the cleanest cut possible. Take only a fine cut each time to prevent tearing the commutator. To finish the job, smooth down the surface with a soft file while the armature is revolving in the lathe between centers. While the armature is still turning, some workers like to polish its surface with various sizes of abrasive paper.

When the armature has been turned, clean between the bars if necessary and then test with a growler or other test instrument to determine if any shorts are present. The vibration of a hacksaw blade on any coil means that the coil is shorted at the leads or commutator. Clean between

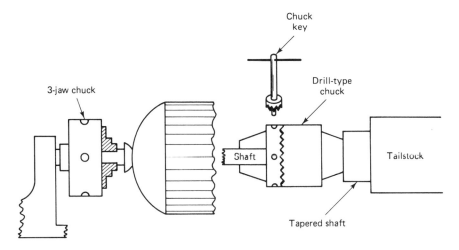

FIG. 14-6 For armatures without center holes in the shaft, lathe chucks must be used to hold the armature while turning.

the commutator bars and test again. As soon as the armature tests okay, it is ready to put back into service.

Motor loads should be reevaluated from time to time, as they will vary for several reasons. Use testing instruments as described in Chapter 17 and take an ampere reading on the motor, first with no load and then at full load – or through an entire cycle. This should give a fair check as to the mechanical condition of the driven machine.

Without proper maintenance, no motor can be expected to perform its best for any length of time or to remain in service as long as it should. Although motor maintenance is costly, it is far less expensive than continually replacing motors or overhauling them frequently.

15

Troubleshooting ac Motors

To detect defects in electric motors, the windings are normally tested for ground faults, opens, shorts, and reverses. The exact method of performing these tests will depend on the type of motor being serviced. However, regardless of the motor type, a knowledge of some important terms is necessary before maintenance personnel can approach their work satisfactorily.

Ground: A winding becomes grounded when it makes an electrical contact with the iron of the motor. The usual causes of grounds include the following: Bolts securing the end plates come into contact with the winding; the wires press against the laminations at the corners of the slots, which is likely to occur if the slot insulation tears or cracks during winding; and the centrifugal switch may be grounded to the end plate.

Open circuit: Loose or dirty connections as well as a broken wire can cause an open circuit in an electric motor.

Shorts: Two or more turns of the coil that contact each other electrically will cause a short circuit. This condition may develop in a new winding if the winding is tight and much pounding is necessary to place the wires in position. In other cases, excessive heat developed from overloads will make the insulation defective and will cause shorts. A short circuit is usually detected by observing smoke from the windings as the motor operates or when the motor draws excessive current at no load.

GROUNDED COILS

The usual effect of one grounded coil in a winding is the repeated blowing of a fuse, or tripping of the circuit breaker, when the line switch is closed, that is, providing the machine frame and the line are both grounded. Two or more grounds will give the same result and will also short out part of the winding in that phase in which the grounds occur. A quick and simple test to determine whether or not a ground exists in the winding can be made with a conventional continuity tester. In testing with such an instrument, first make certain that the line switch is open, causing the motor leads to be *dead*. Place one test lead on the frame of the motor and the other in turn on each of the line wires leading from the motor. If there is a grounded coil at any point in the winding, the lamp of the continuity tester will light, or in the case of a meter, the dial will swing toward *infinity*.

To locate the phase that is grounded, test each phase separately. In a three-phase winding it will be necessary to disconnect the star or delta connections. After the grounded phase is located the pole-group connections in that phase can be disconnected and each group tested separately. When the leads are placed one on the frame and the other on the grounded coil group, the lamp will indicate the ground in this group by again lighting. The stub connections between the coils and this group may then be disconnected and each coil tested separately until the exact coil that is grounded is located.

Sometimes moisture in the insulation around the coils or old and defective insulation will cause a high-resistance ground that is difficult to detect with a test lamp. A megger can be used to detect such faults, but in many cases a megger may not be available. If not, use a test outfit consisting of a headphone set (telephone receiver) and several dry cell batteries connected in series as shown in Fig. 15-1. Such a test set will detect

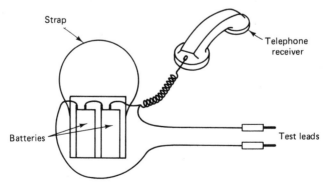

FIG. 15-1   A telephone receiver is very useful for detecting grounds of very high resistance. A clicking sound indicates a fault.

a ground of very high resistance, and this set will often be found very effective when the ordinary test lamp fails to locate the trouble.

Armature windings and the commutator of a motor may be tested for grounds in a similar manner. On some motors, the brush holders are grounded to the end plate. Consequently before the armature is tested for grounds, the brushes must be lifted away from the commutator.

When a grounded coil is located, it should be either removed and reinsulated or cut out of the circuit. At times, however, it may be inconvenient to stop a motor long enough for a complete rewinding or permanent repairs. In such cases, when trouble develops, it is often necessary to make a temporary repair until a later time when the motor may be taken out of service long enough for rewinding or permanent repairs.

To temporarily repair a defective coil, a jumper wire of the same size as that used in the coils is connected to the bottom lead of the coil immediately adjacent to the defective coil and run across to the top lead of the coil on the other side of the defective coil, leaving the defective coil entirely out of the circuit. The defective coil should then be cut at the back of the winding and the leads taped so as not to function when the motor is started again. If the defective coil is grounded, it should also be disconnected from the other coils.

SHORTED COIL

Shorted turns within coils are usually the result of failure of the insulation on the wires. This is frequently caused by the wires being crossed and having excessive pressure applied on the crossed conductors when the coils are being inserted in the slot. Quite often it is caused by using too much force in driving the coils down in the slots. In the case of windings that have been in service for several years, failure of the insulation may be caused by oil, moisture, etc. If a shorted coil is left in a winding, it will usually burn out in a short time and if it is not located and repaired promptly will probably cause a ground and the burning out of a number of other coils.

One inexpensive way of locating a shorted coil is by the use of a growler and a thin piece of steel. Figure 15-2 shows a sketch of a growler in use in a stator. Note that the poles are shaped to fit the curvature of the teeth inside the stator core. The growler should be placed in the core as shown, and the thin piece of steel should be placed the distance of one coil span away from the center of the growler. Then, by moving the growler around the bore of the stator and always keeping the steel strip the same distance away from it, all of the coils can be tested.

If any of the coils has one or more shorted turns, the piece of steel will vibrate very rapidly and cause a loud humming noise. By locating the

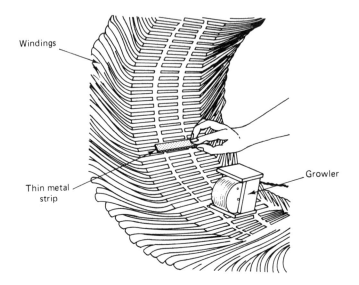

Windings

Growler

Thin metal
strip

FIG. 15-2   Growler used to test a stator of an ac motor.

two slots over which the steel vibrates, both sides of the shorted coil can
be found. If more than two slots cause the steel to vibrate, they should all
be marked, and all shorted coils should be removed and replaced with new
ones or cut out of the circuit as previously described.

Sometimes one coil or a complete coil group becomes short-circuited
at the end connections. The test for this fault is the same as that for a
shorted coil. If all the coils in one group are shorted, it will generally be
indicated by the vibration of the steel strip over several consecutive
slots, corresponding to the number of coils in the group.

The end connections should be carefully examined, and those that
appear to have poor insulation should be moved during the time that the
test is being made. It will often be found that when the shorted end connections are moved during the test the vibration of the steel will stop. If
these ends are reinsulated, the trouble should be eliminated.

OPEN COILS

When one or more coils become open-circuited by a break in the turns or a
poor connection at the end, they can be tested with a continuity tester as
previously explained. If this test is made at the ends of each winding, an
open can be detected by the lamp failing to light. The insulation should
be removed from the pole-group connections, and each group should be
tested separately.

An open circuit in the starting winding may be difficult to locate, since the problem may be in the centrifugal switch as well as the winding itself. If fact, the centrifugal switch is probably more apt to cause trouble than the winding since parts become worn, defective and, more likely, dirty. Insufficient pressure of the rotating part of centrifugal switches against the stationary part will prevent the contacts from closing and thereby produce an open circuit.

If the trouble is a loose connection at the coil ends, it can be repaired by resoldering the splices, but if it is within the coil, the coil should either be replaced or a jumper should be connected around it until a better repair can be made.

REVERSED CONNECTIONS

Reversed coils cause the current to flow through them in the wrong direction. This fault usually manifests itself – as do most irregularities in winding connections – by a disturbance of the magnetic circuit, which results in excessive noise and vibration. The fault can be located by the use of a magnetic compass and some source of low-voltage direct current. This voltage should be adjusted so it will send about one fourth to one sixth of the full-load current through the winding, and the dc leads should be placed on the start and finish of one phase. If the winding is three-phase, star-connected, this would be at the start of one phase and the star point. If the winding is delta-connected, the delta must be disconnected and each phase tested separately.

Place a compass on the inside of the stator and test each of the coil groups in that phase. If the phase is connected correctly, the needle of the compass will reverse definitely as it is moved from one coil group to another. However, if any one of the coils is reversed, the reversed coil will build up a field in the direction opposite to the others, thus causing a neutralizing effect which will be indicated by the compass needle refusing to point definitely to that group. If there are only two coils per group, there will be no indication if one of them is reversed, as that group will be completely neutralized.

When an entire coil group is reversed, it causes the current to flow in the wrong direction in the whole group. The test for this fault is the same as that for reversed coils. The winding should be magnetized with direct current, and when the compass needle is passed around the coil groups, they should indicate alternately N.S., N.S., etc. If one of the groups is reversed, three consecutive groups will be of the same polarity. The remedy for either reversed coil groups or reversed coils is to make a visual check of the connections at that part of the winding, locate the wrong connection, and reconnect it properly.

When the wrong number of coils are connected in two or more

groups, the trouble can be located by counting the number of ends on each group. If any mistakes are found, they should be remedied by reconnecting properly.

REVERSED PHASE

Sometimes in a three-phase winding a complete phase is reversed by either having taken the starts from the wrong coils or by connecting one of the windings in the wrong relation to the others when making the star or delta connections. If the winding is connected delta, disconnect any one of the points where the phases are connected together and pass current through the three windings in series. Place a compass on the inside of the stator and test each coil group by slowly moving the compass one complete revolution around the stator.

The reversals of the needle in moving the compass one revolution around the stator should be three times the number of poles in the winding.

In testing a star-connected winding, connect the three starts together and place them on one dc lead. Then connect the other dc lead and star point, thus passing the current through all three windings in parallel. Test with a compass as explained for the delta winding. The result should then be the same, or the reversals of the needle in making one revolution around the stator should again be three times the number of poles in the winding.

These tests for reversed phases apply to full-pitch windings only. If the winding is fractional pitch, a careful visual check should be made to determine whether there is a reversed phase or mistake in connecting the star or delta connections.

The following troubleshooting chart may be used by qualified personnel who have the proper tools and equipment. These instructions do not cover all details or variations in equipment, nor do they provide for every possible condition to be met in actual practice.

General Troubleshooting Chart

Motor Fails to Start:
1. *Blown fuses.* Replace fuses with proper type and rating.
2. *Overload trips.* Check and reset overload in starter.
3. *Improper power supply.* Check to see that power supplied agrees with motor nameplate and load factor.
4. *Improper line connections.* Check connections with diagram supplied with motor.
5. *Open circuit in winding or control switch.* Indicated by humming sound when switch is closed. Check for loose wiring connections. Also see that all control contacts are closing.

General Troubleshooting Chart (continued)

6. *Mechanical failure.* Check to see if motor and drive turn freely. Check bearings and lubrication.

7. *Short-circuited stator.* Indicated by blown fuses. Motor must be re-wound.

8. *Poor stator coil connection.* Remove end bells and locate with test lamp.

9. *Rotor defective.* Look for broken bars or end rings.

10. *Motor may be overloaded.* Reduce load.

Motor Stalls:

1. *One phase may be open.* Check lines for open phase.

2. *Wrong application.* Change type or size. Consult manufacturer.

3. *Overload motor.* Reduce load.

4. *Low motor voltage.* See that nameplate voltage is maintained. Check connection.

5. *Open circuit.* Fuses blown; check overload relay, stator, and push buttons.

Motor Runs and Then Dies Down:

1. *Power failure.* Check for loose connections to line, to fuses, and to control.

Motor Does Not Come Up to Speed:

1. *Not applied properly.* Consult supplier for proper type.

2. *Voltage too low at motor terminals because of line drop.* Use higher voltage on transformer terminals or reduce load. Check connections. Check conductors for proper size.

3. *Starting load too high.* Check the load that the motor is supposed to carry at start.

4. *Broken rotor bars or loose rotor.* Look for cracks near the rings. A new rotor may be required as repairs are usually temporary.

5. *Open primary circuit.* Locate fault with testing device and repair.

Motor Takes Too Long to Accelerate:

1. *Excess loading.* Reduce load.

2. *Poor circuit.* Check for high resistance.

3. *Defective squirrel cage rotor.* Replace with new rotor.

4. *Applied voltage too low.* Get power company to increase power tap.

Wrong Rotation:

1. *Wrong sequence of phases.* Reverse connections at motor or at switchboards.

Motor Overheats While Running Under Load:

1. *Overloaded.* Reduce load.

2. *Frame or bracket vents may be clogged with dirt and prevent proper ventilation of motor.* Open vent holes and check for a continuous stream of air from the motor.

3. *Motor may have one phase open.* Check to make sure that all leads are well connected.

4. *Grounded coil.* Locate and repair.

5. *Unbalanced terminal voltage.* Check for faulty leads, connections, and transformers.

Motor Vibrates After Corrections Have Been Made:
1. *Motor misaligned.* Realign.
2. *Weak support.* Strengthen base.
3. *Coupling out of balance.* Balance coupling.
4. *Driven equipment unbalanced.* Rebalance driven equipment.
5. *Defective ball bearing.* Replace bearing.
6. *Bearings not in line.* Line up properly.
7. *Balancing weights shifted.* Rebalance motor.
8. *Polyphase motor running single phase.* Check for open circuit.
9. *Excessive end play.* Adjust bearing or add washer.

Unbalanced Line Current on Polyphase Motors During Normal Operation:
1. *Unequal terminal volts.* Check leads and connections.
2. *Single-phase operation.* Check for open contacts.

Scraping Noise:
1. *Fan rubbing air shield.* Remove interference.
2. *Fan striking insulation.* Clear fan.
3. *Loose on bedplate.* Tighten holding bolts.

Noisy Operation:
1. *Air gap not uniform.* Check and correct bracket fits or bearing.
2. *Rotor unbalance.* Rebalance.

Hot Bearings, General:
1. *Bent or sprung shaft.* Straighten or replace shaft.
2. *Excessive belt pull.* Decrease belt tension.
3. *Pulleys too far away.* Move pulley closer to motor bearing.
4. *Pulley diameter too small.* Use larger pulleys.
5. *Misalignment.* Correct by realignment of drive.

Hot Bearings, Ball:
1. *Insufficient grease.* Maintain proper quantity of grease in bearing.
2. *Deterioration of grease or lubricant contaminated.* Remove old grease, wash bearings thoroughly in kerosene, and replace with new grease.
3. *Excess lubricant.* Reduce quantity of grease; bearing should not be more than half filled.
4. *Overloaded bearing.* Check alignment, side, and end thrust.
5. *Broken ball or rough races.* Replace bearing; first clean housing thoroughly.

TROUBLESHOOTING SPLIT-PHASE MOTORS

If a split-phase motor fails to start, the trouble may be due to one or more of the following faults:

1. Tight or "frozen" bearings
2. Worn bearings, allowing the rotor to drag on the stator
3. Bent rotor shaft
4. One or both bearings out of alignment
5. Open circuit in either starting or running windings
6. Defective centrifugal switch
7. Improper connections in either winding
8. Grounds in either winding or both
9. Shorts between the two windings

TIGHT OR WORN BEARINGS.   Tight or worn bearings may be due to the lubricating system failing, or when new bearings are installed, they may run hot if the shaft is not kept well oiled.

If the bearings are worn to such an extent that they allow the rotor to drag on the stator, this will usually prevent the rotor from starting. The inside of the stator laminations will be worn bright where they are rubbed by the rotor. When this condition exists, it can generally be easily detected by close observation of the stator field and rotor surface when the rotor is removed.

BENT SHAFT AND BEARINGS OUT OF LINE.   A bent rotor shaft will usually cause the rotor to bind when in a certain position and then run freely until it comes back to the same position again. An accurate test for a bent shaft can be made by placing the rotor between centers on a lathe and turning the rotor slowly while a tool or marker is held in the tool post close to the surface of the rotor. If the rotor wobbles, it is an indication of a bent shaft.

Bearings out of alignment are usually caused by uneven tightening of the end-shield plates. When placing end shields or brackets on a motor, the bolts should be tightened alternately, first drawing up two bolts which are diametrically opposite. These two should be drawn up only a few turns and the other kept tightened an equal amount all the way around. When the end shields are drawn up as far as possible with the bolts, they should be tapped tightly against the frame with a mallet and the bolts tightened again.

OPEN   CIRCUITS   AND   DEFECTIVE   CENTRIFUGAL SWITCHES.   Open circuits in either the starting or running winding will cause the motor to fail to start. This fault can be detected by testing in series with the start and finish of each winding with a test lamp.

A defective centrifugal switch will often cause considerable trouble that is difficult to locate unless one has good knowledge of the operating characteristics of these switches. If the switch fails to close when the rotor stops, the motor will not start when the line switch is closed. Failure of the switch to close is generally caused by dirt, grit, or some other foreign matter getting into the switch. The switch should be thoroughly cleaned with a degreasing solution such as AWA 1,1,1 and then inspected for weak or broken springs.

If the winding is on the rotor, the brushes sometimes stick in the holders and fail to make good contact with the slip rings. This causes sparking at the brushes. There will probably also be a certain place where the rotor will not start until it is moved far enough for the brush to make contact on the ring. The brush holders should be cleaned and the brushes carefully fitted so they move more freely with a minimum of friction between the brush and the holders. If a centrifugal switch fails to open when the motor is started, the motor will probably growl and continue to run slowly, causing the starting winding to burn out if not promptly disconnected from the line. In most cases, however, the "heaters" in the motor control will take care of this before any serious damage occurs. This fault is likely to be caused by dirt or hardened grease in the switch.

REVERSED CONNECTIONS AND GROUNDS.   Reversed connections are caused by improperly connecting a coil or group of coils. The wrong connections can be found and corrected by making a careful check on the connections and reconnecting those that are found at fault. The test with a dc power source and a compass can also be used for locating reversed coils. Test the starting and running windings separately, exciting only one winding at a time, with direct current. The compass should show alternate poles around the winding.

The operation of a motor that has a ground in the windings will depend on where the ground is and whether or not the frame is grounded. If the frame is grounded, then when the ground occurs in the winding, it will usually blow a fuse or trip the overcurrent device.

A test for grounds can be made with a test lamp or continuity tester. One test lead should be placed on the frame and the other on a lead to the winding. If there is no ground, the lamp will not light, nor will any deflection be present when a meter is used. If the light does light, it indicates a ground due to a defect somewhere in the insulation.

SHORT CIRCUITS.   Short circuits between any two windings can be detected by the use of a test lamp or continuity tester. Place one of the test leads on one wire of the starting winding and the other test lead on the wire of the running winding. If these windings are properly insulated from each other, the lamp should not light. If it does, it is a certain indication that a short exists between the windings. Such a short will usually cause part of the starting winding to burn out. The starting winding is always wound on top of the running winding, so if it becomes burned out due to a defective centrifugal switch or a short circuit, the starting winding can be conveniently removed and replaced without disturbing the running winding.

16

Troubleshooting dc Motors

A new dc motor should always be thoroughly tested before being put into operation and, likewise, should be tested periodically throughout its useful life. By the same token, any motor that is repaired should again be tested before it is put back into service.

GROUND TEST

The fields, armature, and brush holders should be tested for grounds by first disconnecting all external leads. Then use a continuity tester by placing one test lead on the frame of the motor and then touching the other test lead to each motor lead in succession as shown in Fig. 16-1. The tester should not register. If it does, a ground is indicated either in the field circuit or in the armature circuit. Determine which.

Figure 16-2 shows the most likely areas in which a ground will occur. A grounded field coil may be burned and have several wires damaged or broken, requiring a complete rewiring of the field. If a ground is located on the series fields, the interpole, or the shunt fields, it will be necessary to remove the fields from the frame and reinsulate them.

TEST FOR OPENS

The method of testing for opens will vary from motor to motor depending on whether it is series, shunt, or compound wound. In a small series motor, for example, only two leads are used for connection to the line; the

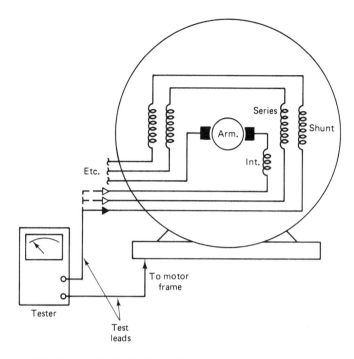

FIG. 16-1   Method of testing a dc compound motor for grounds.

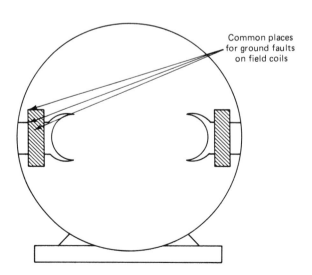

FIG. 16-2   The most likely places where the field will be grounded on a dc motor.

field and armature windings are made internally. When the test leads are connected to these two motor leads, the lamp should light, or the indicator moves on the dial to indicate a complete circuit. If the lamp does not light, the problem could be any of the following:

1. Brushes not making contact with the commutator
2. A broken wire in the field
3. A broken connection between fields
4. A wire disconnected or broken on the brush holder

This same test may be used on large series motors with external leads to field and armature. However, on dc motors with more than one circuit (shunt, two; compound, three) and when only two leads are brought out of the motor, the motor must first be disassembled before the tests can be made.

ARMATURE TESTING

One of the most common devices found in maintenance shops for testing motor armatures is the *growler*, which is constructed of laminated iron in the form of a core around the center of which a coil of insulated wire is wound as shown in Fig. 16-3. When this coil is connected to an alternating current source, it sets up a powerful alternating magnetic field at the two poles of the growler.

Note in the illustration that growlers are made with poles shaped at an angle, so that armatures of different sizes can be laid in these poles. Growlers are also made with poles shaped with the angle in the opposite direction (as shown) so they can be conveniently used on the inside of large alternating current windings.

The growler on the right in Fig. 16-3 has its windings arranged in two separate coils, and the leads are connected to a double-throw, double-pole switch, so that the coils can be used either in series or parallel by changing the position of the switch. This permits the growler to be used on either 120 or 240 V and also makes possible an adjustment of growler field strength for testing windings with different numbers of turns and high or low resistance.

When an armature is placed in a growler and the current turned on in the coil, the flux set up between the poles of the growler builds up and collapses with each alternation, thus cutting across the armature coils and inducing a voltage in them in a manner similar to the action in a transformer. If there are not faults of any kind in the armature winding, no current will flow in the coils from the voltage induced by the growler.

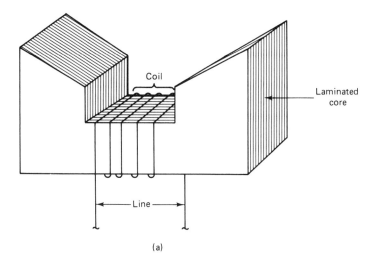

(a)

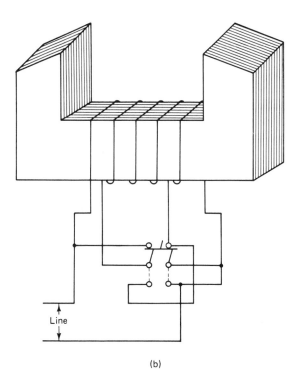

(b)

FIG. 16-3 Two types of growlers for testing motor armatures and stators. The one above (a) is for testing armatures, while the one below (b) is for use inside stator cores.

However, if there is a short circuit between two of the commutator segments or within the turns of a coil, an alternating current will flow in this
shorted coil when it is placed at right angles to the growler flux. This secondary current, which is flowing in the armature coil, will set up alternating flux around it and in the teeth or edges of its slots.

Now if a thin piece of steel is held over the opening of this slot – such
as a piece of hacksaw blade – the steel will vibrate rapidly. A short circuit
is the only fault that will give this indication, so it is a very simple and excellent way for locating shorted armature coils.

It is best to perform all tests with a growler on coils that are in the
same plane of the growler flux; that is, as the test is made from one slot to
the next, the armature should be rotated to make the tests on all coils in
the same position.

An ammeter is sometimes used in conjunction with a growler. The
two leads of the ammeter are placed across a pair of adjacent commutator
bars which connect to a coil lying in the growler flux, and the ammeter
should show a definite reading. If the test is continued around the commutator, testing pairs of adjacent bars while rotating the armature to
make the test on coils which are in the same plane, each pair of bars
should give the same reading. In the case of a faulty coil, the reading may
either increase or decrease, depending on the nature of the fault.

When testing wave-wound armatures, if the leads of two coils are
shorted, the indication will show up at four places around the armature.
Figure 16-4 shows a winding for a four-pole wave armature in position for
testing in a growler. The heavy lines represent two coils which complete a

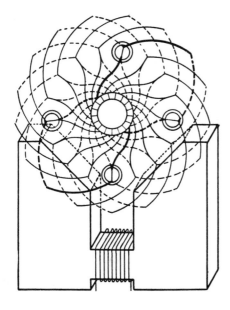

FIG. 16-4   Diagram of coils in a four-
pole, wave-wound armature in place
for testing with a growler. (Courtesy
Page Power Co.)

circuit between adjacent commutator bars, 1 and 2 (see Fig. 16-5). The top side of one of these coils and the bottom side of the other connect at bar 10. Note that a short circuit between bars 1 and 2 would cause the steel strip to vibrate over the four slots shown by the small double circles.

In addition to short circuits, a number of other problems may occur in dc motors. They include grounded coils or commutator bars, open coils, shorts between commutator bars, and reversed coil leads. In addition to the growler, which can be used to locate any of these faults, a galvanometer and battery may be used to locate many of these troubles by testing at the commutator bars.

Figure 16-5 shows a diagram of a two-pole winding with a number of the more common faults which may occur in armature coils and at the commutator segments. In reviewing this diagram, note that coil 1 is shorted within the coil, which is probably the result of broken or damaged insulation on the conductors. To test for this fault, place the armature on the growler and close the switch to excite the growler coil. Place a thin steel strip over an armature slot which is at least the distance of one coil span from the center of the growler core. Turn the armature slowly, keeping the steel parallel with and over the slots. When the slot containing coil 1 is brought under the steel, the induced current flowing in this

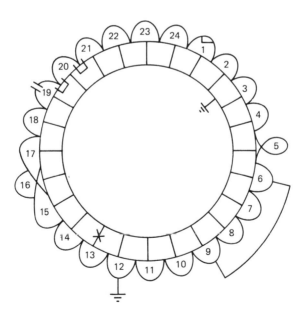

FIG. 16-5 Diagram showing a two-pole, lap-wound motor with a number of common faults which occur in the armature coils and at the commutator segments.

local short circuit will set up flux between the teeth of this slot, which will attract and repel the steel strip, causing it to vibrate like a buzzer. This indicates that the coil is short-circuited. Mark this slot with a piece of chalk and proceed with the test. Again rotate the armature slowly and test each slot, at all times keeping the strip over slots that are in the same position with respect to the growler. When the slot which contains the other side of the shorted coil is brought under the steel strip, it will again vibrate. Mark this slot also. The two marked slots should now show the span of the exact coil which is shorted.

If no other slots cause the steel to vibrate during the test, there is only one short in the armature. This test will apply to armatures of any size, regardless of the number of poles in their winding and whether they are wound lap or wave.

To locate the bars on the commutator to which the leads of the shorted coils are attached, use the ammeter, placing the two test leads on two adjacent bars. Note the reading on the meter carefully and, by rotating the armature, check the readings of all the other bars in this same position. When the test leads are placed on the bars that connect to the shorted coil, the reading will be lower than the other readings obtained.

In testing for loose coil leads, such as the ones shown on coils 20 and 21 of Fig. 16-5, the growler and thin strip will not show any vibration at any slot. However, the ammeter may be used (with the growler) by testing between commutator bars. When the ammeter leads are placed on the commutator bars to which these coils are connected, the reading between them and adjacent bars would drop to zero, indicating an open circuit.

The open circuit shown in coil 19 in Fig. 16-5 can be found by testing around the commutator with an ammeter. When the test leads of the ammeter are placed across the bars to which the open coil is connected, a very low reading will be obtained. The reason for any reading at all is because there are always two paths for the current to travel through the winding, unless it is open at some other coil also. With an open circuit only at coil 19 in the diagram, there would still be a circuit through all the other coils in series. The voltages induced in the coils which lie in the active position for the growler flux would tend to neutralize each other, but there is often a slightly unbalanced condition in the windings which would allow a little current to flow through the ammeter.

If there are, say, three coils of the armature in the active flux of the growler and one side of coil 19 (see Fig. 16-5) is one of these, then there will be three good coil sides working against two good coil sides with their induced voltages, and since coil 19 is open-circuited, the ammeter reading would be about one third normal. The exact amount of the reading, however, will depend on the pitch of the coils and the size of the armature. Be sure to remember that one open circuit in an armature does not necessarily give a zero reading on the ammeter unless the coil sides on each side of the test points are perfectly balanced electrically.

REVERSED COIL

Coil 5 in Fig. 16-5 is reversed. Neither the growler with the steel strip nor the growler with the ammeter will show this fault, because the induced current is alternating and the motor will not indicate the reversed polarity of the coil. However, if the test leads of the ammeter are spread far enough apart to touch bars 1 and 3, two coils in series will be tested. Then place the test leads on bars connected to coils 4 and 5 or 9 and 6 (two coils will be in series in either case), and it can be found that the voltage in one will be opposite in direction to that in the other, causing the reading to be zero. To summarize, in testing for reversed coils, the test leads are placed on the proper bars to test an extra commutator segment, and the indication for the reversed coil will be a zero reading.

GROUNDED COILS

Coil 12 in Fig. 16-5 is grounded, but the growler and steel strip test, or the bar-to-bar test with the ammeter, will not indicate this fault. To locate a ground, the test leads are placed one on the commutator and one on the shaft or core of the armature. If the first test is made between the bar of coil 8 and the shaft, a very high reading would be obtained on the ammeter, because this would give the reading of the four coils in series between the grounded coil and this bar.

As bars are tested closer to the grounded point, the reading will gradually decrease, and the two bars that give the lowest reading should be the ones connected to the grounded coil. The sum of the readings from these two bars to the shaft should equal the reading of a normal coil.

SHORTS BETWEEN COILS

In referring again to Fig. 16-5, it is found that coils 6 and 9 are shorted together, which places coils 6, 7, 8, and 9 in a closed circuit through the short and the coil connections to the commutator bars. In this case, the growler and steel strip will vibrate and indicate a short circuit over each of the slots in which these coils lay. A bar-to-bar test with the ammeter leads would not give a definite indication, but the readings on these bars would be lower than normal.

REVERSED LOOPS

Coils 15, 16, and 17 in Fig. 16-5 are connected properly to each other, but their leads are connected to the wrong commutator bar. To detect this, use the bar-to-bar test with an ammeter. In doing so, the meter will show double readings between bars 1 and 2, normal readings on bars 2 and 3, and double readings again on bars 3 and 4. This indicates that the coils are connected in proper relation to each other but that their leads are crossed at the commutator bars.

SHORTED COMMUTATOR SEGMENTS

Coil 13 in Fig. 16-5 has a short between its commutator bars. To detect this fault, the growler and thin steel strip may be used, in which case the steel strip would vibrate, indicating a short circuit over both slots in which this coil lies. The bar-to-bar test of the ammeter will give a zero or very low reading across these two bars, depending on the resistance of the short circuit between them.

If the winding is connected lap, the short would be indicated in two places on the core, and if it is connected wave for four poles, it would be indicated in four places on the core.

GROUNDED COMMUTATOR SEGMENTS

Assume that the commutator bar to which coils 2 and 3 are connected in Fig. 16-5 is grounded to the shaft. The growler and steel strip will not indicate this fault. The ammeter used in conjunction with the growler, with tests made between other commutator bars and the shaft, would show high readings on the meter, but as bars are tested closer to the grounded one, the reading falls lower and lower and will be zero when one test lead is on the grounded bar and the other on the shaft. If an absolute zero reading is obtained, it indicates the ground is at the commutator bar.

GALVANOMETER TESTS ON ARMATURES

A galvanometer is a measuring device for indicating very small electric currents. The D'Arsonval galvanometer is the most common type and is widely used in the electrical industry. Its indicating system consists of a light coil of wire suspended from a copper ribbon a few thousands of an inch wide and less than .001 in. thick. This coil, free to rotate between the poles of a permanent magnet, carries a small mirror which serves as an optical pointer and indicates the coil position by reflecting a light beam onto a fixed scale. The torque which deflects the indicating element is produced by the reaction of the coil current with the magnetic field in which it is suspended.

Figure 16-6 shows a method of making galvanometer tests on armatures. Two leads from a dc power source should be held against bars on opposite sides of the commutator and kept in this position as the armature is rotated. This will send a small amount of direct current through the coils of the winding in two paths in parallel.

If the positive lead in Fig. 16-6 is on the right, a current will flow from this lead through the commutator bar to the right-hand side of the winding. If all coils of the winding were closed and in good condition, the current would divide equally, part flowing through the top section of the winding to bar 3 and the negative lead and the other part flowing through the lower section of the winding to the same bar and lead. When this current is flowing through the armature and a test is made between adjacent bars with the galvanometer, the instrument reads the voltage drop due to the current flowing through the resistance of each coil. So the galvanometer test performs quite similarly to the test with the ammeter and growler.

In testing for an open circuit with the galvanometer leads placed on adjacent bars connected to good coils, there will be no reading in the section of the winding in which the open coil is located, but when these leads are placed across the bars connected to the open coils, the needle will probably jump clear across the scale, because at this point it tends to read practically the full dc voltage.

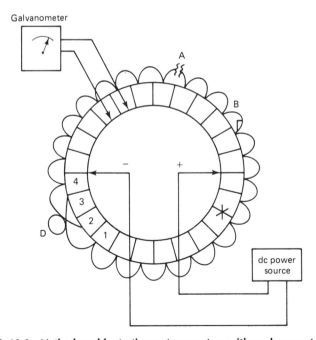

FIG. 16-6 Method used for testing motor armature with a galvanometer.

Of course, if there are two open circuits in this half of the armature, no reading will be obtained at any pair of bars. This is a good indication that more than one open is present. If a test is made all the way around the commutator and no open circuits are present, the galvanometer should read the same across any pair of bars. You should be careful, however, to secure at all times a good contact between these test leads and the bars and also be certain that the battery or dc voltage leads made a good connection to the commutator as the armature is rotated. Otherwise, variations in the readings will be obtained.

A reading lower than normal between any two bars will indicate a shorted coil, and a zero reading indicates a short between two commutator bars. When galvanometer leads are placed on bars 2 and 3, which are connected to coils with their leads transposed, the reading will be normal, but in testing between bars 1 and 2 or 3 and 4 the reading will be double. This indicates that the leads at bars 2 and 3 are the ones reversed.

REMOVING FAULTY COILS

In many cases when a motor develops problems, it is inconvenient to take it out of service for complete rewinding or for the amount of time required to replace the defective coils with new ones. At times like these, when it is extremely important that a machine be kept in service to avoid delays in production, a quick temporary repair can be made by cutting the faulty coils out of the armature circuit. This is done by using a jumper wire of the same size as the conductors in the coils and which should be soldered to the same two bars to which the defective coil was connected. This jumper will then complete the circuit through this section of the armature and will carry the current that would normally have been carried by the defective coil.

Figure 16-7 shows the manner in which an open-circuit coil can be cut out with such a jumper. For each coil that is cut out of the winding, a slightly higher current will flow through the other coils of that circuit. The number of coils that can safely be cut out will depend on the position in which they occur in the armature.

In some cases, several coils may be cut out if they are equally distributed around the windings, but if several successive coils became defective and were all cut out with a jumper, it might cause the rest of the coils in that circuit to burn out.

Other factors that determine the number of coils which can be cut out in this manner are the following:

1. The number of coils per circuit
2. The amount of load on the motor or generator
3. The size of the machine

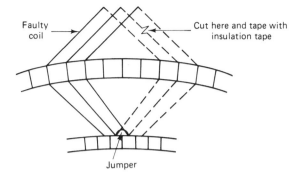

Faulty coil

Cut here and tape with insulation tape

Jumper

FIG. 16-7 Method of cutting out a defective coil and completing the circuit through the winding with a jumper at the commutator bars to which the defective coil connects.

If the defective coil is grounded, its two ends should be disconnected from the commutator bars before the jumper is soldered in place. Shorted coils should be cut at the back end of the armature and these cut ends well insulated. The jumper wire should be well insulated from the leads of other coils.

CONSTRUCTING GROWLERS

We have mentioned the use of a growler in this chapter for all types of troubleshooting applications. While growlers are commercially available from motor repair suppliers, many maintenance shops prefer to build their own.

Laminations designed for use in making small transformers may be used to construct a growler for use in testing motor armatures or stators. The drawings in Fig. 16-8 show how the laminations may be trimmed and arranged for use in constructing a growler.

After the laminations are trimmed as shown by the dotted lines [Fig. 16-8(a)], they are stacked as shown in Fig. 16-8 (b), forming the letter H. Place the piece with the center bar attached on the work bench and then butt the I piece against the center bar as shown. The next two laminations are reversed so as to break joints. In other words, if the I piece is on the right for the first layer, it should be on the left for the next layer, and so on. Continue stacking the laminations alternately on first one side and then the other until you have a stack about 1 in. high.

The laminations must now be bound together either with bolts or by using a bar clamp arrangement as shown in Fig. 16-8(c). Two pieces of fiber or wood about 3 in. long with a hole in each end may be used as a clamp. After the core is assembled, it should be carefully insulated. The part of the core which will come in contact with the wire should be

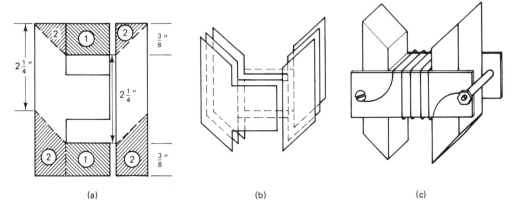

FIG. 16-8 Steps in making a growler for testing motor armatures and stators.

covered with a layer of varnished cambric or other insulating paper which may be wound around the core and over the fiber strips.

The winding of the growler consists of approximately 2000 turns of No. 34 AWG insulated wire. This wire should be carefully wound on the center part of the core as shown in Fig. 16-8(c). Terminals may be provided on the insulated clamp so that the ends of the coil may be attached to them, or the two clamping bolts may be used for the terminals.

Once the construction of the coil is completed, dip it in insulating varnish or else wind it with tape to protect the coil. This particular size is suited for testing small to medium armatures or stators. Other sizes may be constructed using the same principle.

17

Testing and Measuring Instruments

When using any instrument for testing or measuring electrical circuits, always consider your personal safety first. Know the voltage levels and shock hazards related to all wiring and equipment to be tested. Also be certain that the instrument used for the application has been tested and calibrated.

When taking readings with meters available with different ranges and functions – as in the case of a combination volt-ohm-ammeter – make certain that the meter selector and range switches are in the correct position for the circuit to be tested.

If possible, the circuit being tested should be deenergized before attaching the meter or test leads to the desired test point. The circuit may then be energized to take the reading. Before disconnecting the meter, the circuit again should be deenergized.

VOLT-OHM-AMMETER

The volt-ohm-ammeter is one of the most useful testing instruments that the maintenance electrician can use (see Fig. 17-1). Some of its uses are as follows.

MEASURING INSULATION RESISTANCE. On partial grounds in motor windings, the insulation resistance can be measured with a voltmeter if the resistance of the voltmeter is known. The deflec-

171

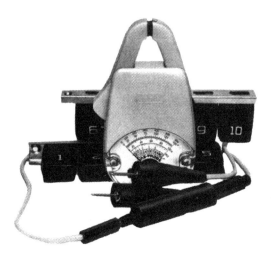

FIG. 17-1 The Amprobe Line-Probe is a compact measuring device that is suitable for all kinds of tests in electrical work, including motor trouble-shooting and repairs. (Courtesy Amprobe Instrument, Division of Core Industries, Inc.)

tion of a voltmeter pointer is inversely proportional to the resistance of its circuit. Therefore, if E is the deflection measured directly across the circuit, E' the deflection measured through the insulation resistance, r the resistance of the voltmeter, and R the resistance of the insulation, then

$$\frac{E'}{E} = \frac{r}{R+r} \quad \text{or} \quad R = r\left(\frac{E}{E'} - 1\right)$$

The deflection measured directly across the circuit and also through the insulation resistance will be in volts, while r and R will be in ohms.

DETERMINING FAULTS IN FIELD CIRCUITS. To find an open-circuit field coil in an electric motor, first connect a source of voltage across the terminals of the field circuit as shown in Fig. 17-2 at a and e and apply the leads of a voltmeter successively to the field-coil terminals (ab, be, etc.). No deflection of the voltmeter pointer will occur until the leads are applied to the terminals of the defective coil, and then the deflection will show the voltage between a and e.

Another method is to leave one lead of the voltmeter connected to one side of the circuit as at a and the other touched to successive coil terminals b, c, etc. No deflection will occur until the defect is bridged, and the voltmeter will then show the full voltage between a and e.

SHORT-CIRCUITED FIELD COIL. To find a short-circuited field coil, proceed as described for finding an open circuit but limit the current through the field circuit to a safe amount. The drop across the defective field coil, as indicated by the voltmeter, will be much less than that across any other coil.

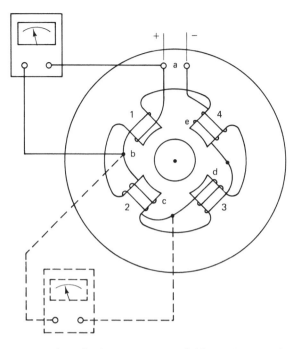

FIG. 17-2 Connection of voltmeter to motor field circuit terminals to find an open circuit.

FAULTS IN MOTOR ARMATURES. In winding an armature, it is advisable to test the winding for grounds and short circuits before the coils are connected to the commutator; any defective coils can then be replaced. The armature can be given an insulation-breakdown test by joining all the free ends of the coils together by means of a small wire and then connecting the series of coils to one terminal of a high-voltage transformer. The other terminal of the transformer is connected to the motor's shaft. The alternating current to be applied for breakdown tests on completed motors is given in the following table:

Rated Voltage of Motor Under Test	Testing Voltage
Not exceeding 400 V	1000 V
Between 400 and 800 V	1500 V
Between 800 and 1200 V	3500 V
Between 1200 and 2500 V	5000 V
Between 2500 and 10,000 V	Double rated voltage
10,000 V and over	10,000 V above normal

SHORT-CIRCUITED ARMATURE COIL. A short-circuited armature coil can be found by sending through it an alternating flux, which will set up a current in the coil. Special testing transformers are normally used for this test. The transformer consists of a laminated iron core which has a width approximately equal to the length of the armatures to be tested and pole-face curvature about the same as that of the armature cores.

When the transformer is placed over an armature core and an alternating current is sent through the magnetizing coil, the alternating flux induces a voltage in the armature coils under the transformer poles; if any coil is short-circuited, the resulting current will heat it or will cause a piece of iron held near it to vibrate. By turning the armature through a complete revolution and holding a piece of iron over the coils being tested, any short-circuited coils are easily located. The transformer would not be put in position or removed while current is flowing through the magnetizing coil.

BAR-TO-BAR ARMATURE TEST.   This test will reveal the existence and location of open circuits, short circuits, incorrect connections, and grounds in a completed armature. A steady current, adjusted to suit the requirements, is sent through the armature by terminals clamped to opposite sides of the commutator, and the leads of a voltmeter are connected adjacent to the commutator segments by means of contact points insulated from each other and held a fixed distance apart by a block of wood or other insulation material. If there are no defects in the armature, the reading of the voltmeter will be the same for each pair of adjacent segments, because the drop of voltage is the same in all the coils. By trying several pairs, one can obtain a standard deflection, adjusting the current until this deflection is easily readable. If two bars are short-circuited or a coil is short-circuited, the voltmeter will show almost no deflection when the contacts are on the bars 1 and 3, 2 and 4, etc. A poor connection between a bar and the two coil leads belonging in it will be revealed by an unusually large drop between the bar and either adjacent bar. An open circuit in any coil prevents flows of current through that portion of the armature, and when the points bridge any two bars in the open section, the voltmeter will show no deflection until bars 9 and 10 (Fig. 17-3) are bridged, when the full voltage between terminals *a* and *b* will be impressed on the instrument and the needle will be thrown violently. Crossed leads will cause double deflection when the points are on bars 6 and 7 or 8 and 9, because the voltage drop is then measured in two coils; between bars 7 and 8 the drop is normal in value but reversed in direction.

GROUNDED ARMATURE COIL.   A grounded armature coil can be found by impressing full voltage on the test brushes, connecting one

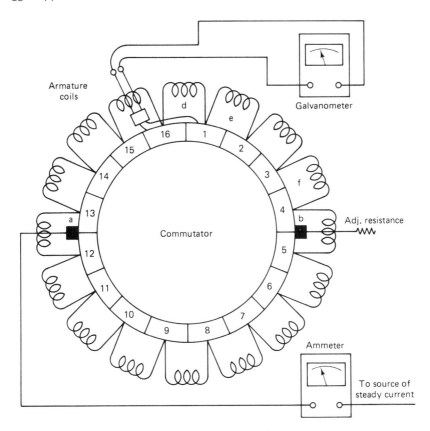

FIG. 17-3   Test connection for bar-to-bar armature test.

voltmeter lead to the shaft, and touching the other successively to each commutator bar. The deflection will be least when a bar to which the grounded coil is connected is touched.

MEGGER APPLICATIONS

The megger insulation tester is a practical instrument used by electrical maintenance and repair workers for conducting a wide range of tests with dependable accuracy. It has earned the enviable distinction of being the most widely accepted device for measuring electrical resistance. For this reason every electrical worker should have a good knowledge of the megger's operation and the practical application of the meter.

One use of the megger is for testing electrical systems after they are installed and before normal voltage is applied. This test is made after all the conductors, fuses or circuit breakers, panelboards, outlets, etc., are in place and connected. The current used for testing is produced by a small generator within the megger that generates dc power by turning a crank handle (also a part of the megger).

The test is made by connecting the terminals to the two points between which the insulation test is to be made and then rapidly turning the handle (crank) on the megger. The resistence in megohms can then be read from the dial. Satisfactory insulation resistance values will vary under different conditions, and the charts supplied with the megger should be consulted for the proper value for a particular installation.

A similar test is often made with a magneto or small generator in conjunction with a dc-operated bell. However, this is not nearly as reliable as the megger since it does not give any indication of the insulation resistance on a dial but merely rings the bell if there is any trouble on the lines tested.

Most modern meggers have an additional terminal which is referred to as a *guard* terminal. This guard system is a means of preventing errors in the ohmmeter readings due to current leakage inside or on the outside of the instrument between the positive and negative sides of the circuit. Such leakage may be caused by dampness or dirt.

Figure 17-4 shows how the guard system is used. The guard ring is a

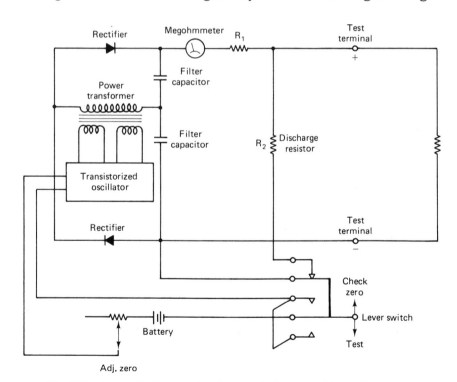

FIG. 17-4 Schematic diagram showing how leakage guard is used on megger testing instrument.

metallic washer supporting the line terminal and insulated from it. Any leakage current that may creep across the surface of the case or through materials from the positive ground (earth) terminal toward the negative line terminal will be intercepted by the guard ring. The guard circuit offers a low-resistance path for the leakage current directly to the dc source without passing through the deflecting coil of the ohmmeter. Resistance coils and other live parts inside the instrument are also mounted on guarded supports.

GROWLER FOR TESTING ARMATURES

The growler has been around for a long time and is still a very good testing device for use on electric motors. The growler operates on the same principle as a conventional transformer and is made up of a number of laminations secured together so a coil may be wound around the laminated body of the growler. The growler acts as a primary side of the *transformer*, while the armature or stator coil acts as the secondary side. If there is a short in a secondary side, it will set up currents which will vibrate a thin piece of metal such as a section of hacksaw blade. A growler is always used on alternating current — never on direct current.

To use the growler, lay the armature on the growler as shown in Fig. 17-5. Turn the armature around slowly in the growler and lay a section of hacksaw blade over each coil. A short-circuited coil will cause the blade to

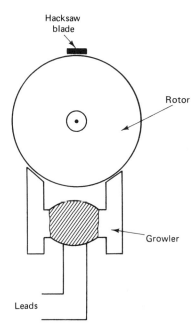

**FIG. 17-5 Use of a growler for testing   Leads
motor armature.**

vibrate. If any coil should get hot, then that coil is shorted. In fact, defective coils will normally get hot after being placed in the growler for only a minute or so.

When the blade is placed so as to make contact with the commutator bars and shaft, a spark will usually exist if there is a ground.

On open and closed circuits, the armature is laid on the growler, and a hacksaw blade is placed on the bars so as to make contact with two bars together. Somewhere on two or more bars, the test should cause a spark when the contact of the blade is made between them. This means the circuit is closed. If there are no sparks, the armature has an open circuit.

In using the growler to make the various tests, it is best to lightly scrape dirt from between commutator bars before starting any tests, assuring that a true test will result. Greasy commutators should be washed with gasoline and blown off or allowed to dry before performing the test.

For the price, the growler is one of the most useful test devices for troubleshooting motors of all kinds. Most maintenance shops have the tools and materials to construct any size growler necessary in a very short period of time. See Chapter 16 for complete details on constructing a growler for testing motor armatures or stators.

SPECIAL TESTERS

In recent years, many types of special motor testing instruments have become available. One is the Motor Tester PMZ as shown in Fig. 17-6. Tests of all types can be made with this instrument, and it is available from Dreisilker Electric Motors, P.O. Box 2758, Glen Ellyn, Illinois 60137.

Direct testing at the terminals and magnetic testing with test probes make this instrument suitable for detecting all motor deficiencies, including opens, short circuits, grounds, and an incorrect number of turns.

For direct measuring, the PMZ is connected with the main circuit connections of the test object. The state of the motor can be recognized before it has been detached. Faults on the terminals of field coils, carbon brushes, and collector segments can be easily determined. Test prods or a contact fork for the collector segments allow a quick contact.

Probes allow the testing of disassembled armatures or stators after the removal of the rotor by a quick scanning. Each fault can be exactly localized for each slot. The single probe 8 can be used for stators with a minimum hold 25 mm in diameter and a height of the bundle of laminations from 20 mm and for large motors with a slot up to 8 mm.

For direct measurement, the ac resistance of the stator or armature

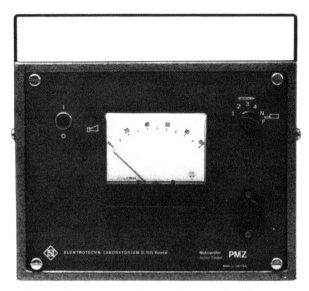

FIG. 17-6 An excellent tester for shops doing considerable motor testing and repair is the PMZ motor tester. (Courtesy Dreisilker Electric Motors, Inc.)

is measured at 2 kHz, thus allowing easy determination of faults and damages.

During the measurement with probes, the probe generates a magnetic field in the test object and takes up the reaction of faulty windings. The measuring instrument indicates exactly, and the built-in buzzer announces faults acoustically.

INSULATION BREAKDOWN TESTER UHZ

The insulation tester in Fig. 17-7 is capable of testing the insulation strength of electric motors and other equipment with high alternating voltage, offering a sine-shaped ac voltage steplessly adjustable from 0 to 2000 V.

A key lock prevents unauthorized use, and the high voltage potential is insulated from the ac line and isolated to ground, making the UHZ qualified for both permanent and field test areas.

The test voltage is attached to the test object by two hand-held, high-voltage, safety test probes. For tests requiring more than 2000 V, use an insulating safety cage or an automatic test system.

It is possible to switch the UHZ from *test* to *burn*. In the test position, the high tension will be switched off within a few cycles of the mains if the insulation of the test piece breaks down. A buzzer indicates the

FIG. 17-7   The insulation breakdown tester UHZ is capable of testing the insulation strength on all types of electrical apparatus. (Courtesy Dreisilker Electric Motors, Inc.)

fault. To locate the fault, switch the UHZ to the burn position and cutoff is delayed.

TESTING OF STEP MOTORS

Step motors are devices which translate electrical pulses into mechanical movements. The output shaft rotates or moves through a specific angular rotation per each incoming pulse or excitation. This angle or displacement per movement is repeated precisely with each succeeding pulse translated by appropriate drive circuitry. The result of this precise, fixed, and repeatable movement is the ability to position accurately.

The testing of step motors can involve a variable assortment of tests and techniques. The wide variety of step motor applications and the corresponding variation in the tasks which are performed by them indicate that a particular motor will have to be tested based on its function in the system.

When used with phase-switched inputs, a step motor can be a very nonlinear device. Its behavior may therefore vary under different load and drive configurations. The tests performed with a step motor should, insofar as possible, represent the actual conditions under which it is to be used. This is particularly true with open-loop tests when load configuration and input pulse sequences can considerably affect the presence and location of resonant conditions.

Frequent testing is required by the user; this can include any combination of the following tests, depending on the application:

1. Torque-speed characteristics with a specific drive scheme
2. Static torque characteristic as a function of position
3. Single step response
4. Static positional accuracy test
5. Velocity measurements
6. Closed-loop switch angle vs. speed measurements
7. Inductance measurement

In the open-loop mode this test generally involves driving the motor up to a particular speed and loading it slowly while continuously measuring the torque until the motor is pulled out of synchronization. Due to the nature of step motors in the open-loop mode, the results of a torque speed test will, in general, depend on the drive scheme, the type of load, and the type of torque-measuring instrument.

CORD AND SPRING SCALE.   In this system a cord wrapped on a pulley is used to apply the frictional torque, and a spring scale is used for measurement. Figure 17-8 shows a schematic arrangement of this technique.

HYSTERESIS DYNAMOMETERS.   This method uses a hysteresis brake for applying the frictional load and a gravity balance for measurement of the torque. A commonly available unit from Magtrol, Inc., is

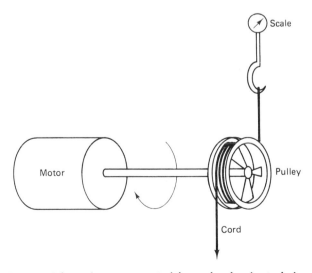

FIG. 17-8   Schematic arrangement of the cord and spring technique.

shown in Fig. 17-9. Although this unit shows a calibrated scale for torque readout, the newer units are available with a direct reading digital meter.

MAGNETIC PARTICLE BRAKE DYNAMOMETERS.   A magnetic particle brake can provide a uniform, low-inertia load, which is adjustable with input current. The brake consists of a rotor surrounded by fine magnetic particles and enclosed in a nonrotatic housing. A coil, when energized, creates a magnetic field, which tends to solidify the particles, thus loading the rotor. The load torque is controlled by current in the coil.

An optical torque-bar transducer provides an efficient torque-measuring device for use with the particle brake. Figure 17-10 shows the principle of such a device available from Vibrac Corporation. An analog or digital readout is available which directly displays the torque.

DC MOTOR DYNAMOMETERS.   The use of modern low-inertia servo motors and controls can provide an effective steady-state torque-measuring system. The basic advantage of this type of dynamometer is that it can provide four-quadrant operation, and retarding torques can be measured.

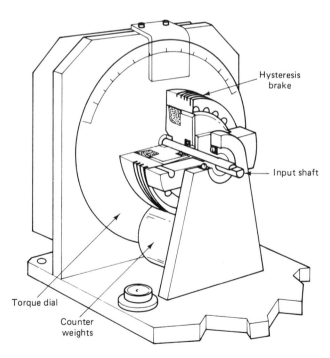

FIG. 17-9   Magtrol, Inc., hysteresis dynamometer. (Courtesy The Superior Electric Company.)

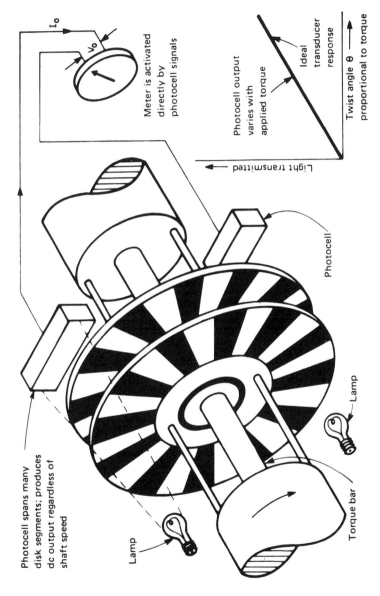

Meter is activated directly by photocell signals

I_o

V_o

Photocell output varies with applied torque

Ideal transducer response

Light transmitted

Twist angle θ ——— proportional to torque

Photocell spans many disk segments; produces dc output regardless of shaft speed

Photocell

Lamp

Lamp

Torque bar

FIG. 17-10 Principle of an optical torque-bar transducer. (Courtesy The Superior Electric Company.)

Typical torque data obtained with the first three types of dynamometers are shown in Fig. 17-11. The motor is an M093FC11 step motor run with the split resistor scheme, 24 V, 5 A/phase. The comments made earlier regarding each of these types of dynamometers appear to be depicted in these curves.

For closed-loop torque speed testing, the choice of dynamometers is less critical as the motor is inherently more stable and the various resonances are less predominant.

STATIC TORQUE MEASUREMENT. The static torque developed by a step motor varies with rotor position and stator current. The positional variation is cyclic and has a period in steps equal to the number of phases. Figure 17-12 shows a typical static torque curve for a 15° permanent-magnet step motor. The curves for two adjacent phases are shown when energized separately, and a third curve when both phases are energized together is also shown.

When the peak static torque is desired, it is possible to energize the motor and use a simple torque watch or a lever arm to get the data. However, if the complete torque curve is desired, it is necessary to provide a fixture for slowly rotating the motor while it is energized and coupling it

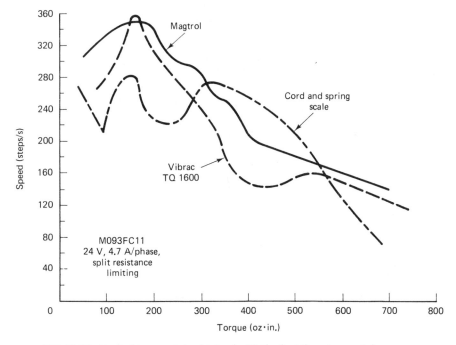

FIG. 17-11 Typical torque data obtained with the first three types of dynamometers.

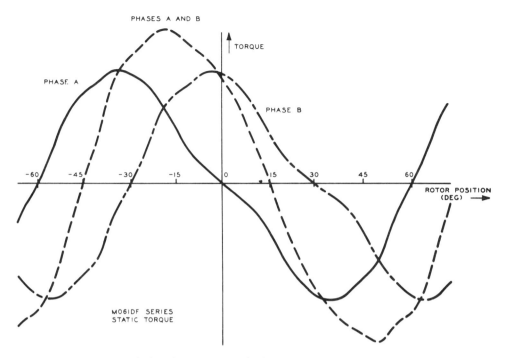

**FIG. 17-12   Typical static torque curve for a 15° permanent-magnet step motor.
(Courtesy The Superior Electric Company.)**

to a potentiometer and a high-stiffness, torque-bar-type transducer. The optical unit from Vibrac is adequate.

Figure 17-13 shows a schematic of such a setup. The rotating assembly should be geared for a slow-speed, low-backlash operation to allow proper data collections in the unstable portion of the torque characteristic. An X-Y plotter is convenient for directly plotting the characteristic.

STEP MOTOR POSITIONAL MEASUREMENT.   The measurement of step motor position can be done with digital or analog transducers. For general response purposes a continuous rotation potentiometer is quite adequate. This device should have a carbon film or conductive plastic element. The inertia should be low compared to the motor rotor inertia.

Continuous rotation potentiometers are useful for single-step or multistep responses as well as for plotting torque vs. position characteristics. They are indispensable when damping studies are performed.

A voltage biasing arrangement using a second potentiometer as shown in Fig. 17-14 is useful for measuring small changes on an oscilloscope without a large voltage bias. It is also helpful for noise reduction to

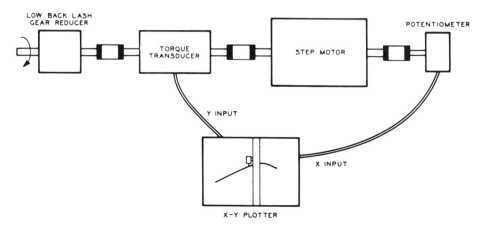

FIG. 17-13 If the complete torque measurement curve is desired, it is necessary to provide a fixture for slowly rotating the motor while it is energized and coupling it to a potentiometer and a high-stiffness, torque-bar-type transducer. (Courtesy The Superior Electric Company.)

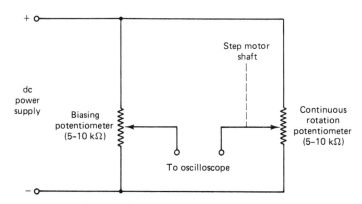

FIG. 17-14 Voltage biasing arrangement using a second potentiometer.

use a separate power supply or a battery for powering the potentiometers.

STEP MOTOR VELOCITY MEASUREMENTS. Velocity information for step motors can be obtained directly with dc tachometers or derived from a high-resolution encoder. Due to its ease of operation and utilization, the dc tachometer is by far the more popular device. The voltage output of a dc tachometer varies linearly with speed. This device is capable of giving transient as well as steady-state velocity measurements.

A dc tachometer can be of the iron core type or moving coil type. The iron core unit will have a ripple of 3–5%, while the moving coil type can have a ripple of less than 1%. An RC filter (typically 1 μF, 10 kΩ) is used to eliminate high-frequency noise. The frequency of the ripple depends on the number of commutator segments and is generally of the order of 7, 11, or 15 cycles per revolution. When a tachometer is used for measuring velocity variations on a motor shaft, care must be taken to account for the tachometer ripple.

TACHOMETER

Electrical tachometers are speed indicators used to indicate or record the speed of a machine or motor. The scales may be calibrated to read directly in revolutions per minute or feet per minute of any driven mechanism, such as a conveyor or motor drive. Tachometer scales may also be calibrated to indicate miles per hour, as in locomotive or railway speedometers.

An electrical tachometer consists of a small generator belted or geared to the unit whose speed is to be measured. The voltage produced in the generator varies directly as the speed of the rotating part of the generator. Since this speed is directly proportional to the speed of the machine under test, the amount of the generated voltage is a measure of the speed. The generator is electrically connected to an indicating or recording instrument which is calibrated to indicate units of speed. An electrical tachometer is particularly well suited to the remote indication of the speed of a unit.

18

Tools for Motor Repair

Good work in any trade requires good tools and, importantly, the knowledge of how each is used. The motor repair field is no exception. Electricians and maintenance personnel should be selective in choosing their tools, know how each is used, know how to care for them, and put them to their best use. Besides the conventional hand tools, there are many specialty tools that will make certain repair techniques easier to accomplish. Also, the host of power tools available on today's market will help to accomplish the various tasks more quickly and usually in a more efficient manner.

This chapter is designed to describe the more common tools used in motor repair—including hand tools (usually purchased by repair workers), specialty tools (some supplied by the plant, some purchased by repair workers), and power tools, which are almost always furnished by the place of employment. A careful review of these various tools will help the beginner or apprentice make the correct initial selection to act as a foundation on which to build a good assortment of tools necessary for motor repair. Even the experienced journeyman may find new and useful data on the use of tools to make jobs easier and work more efficient. Finally, supervisory personnel will have a good knowledge as to what tools may be necessary in the maintenance shop for motor repair and related situations.

While many major motor repair jobs have gone to automatic machines for their overhaul projects, there are thousands of smaller motor repair shops—including maintenance shops in most industrial plants—

that do not require such elaborate equipment. These smaller shops must stick to traditional tools and methods, utilizing the less expensive power tools so that the shops may reach their top efficiency.

To test and rewind motors efficiently, certain tools and testing equipment are necessary. Chapter 17 covered many testing devices that are suitable for use on electric motors. The list of tools that follow should be considered the minimum for each motor repairman. More may be added as the need arises.

---

1 16-oz machinist's hammer
1 12-oz machinist's hammer
1 Large screwdriver, 6 in.
1 Small screwdriver, 3 in.
1 #1 rawhide mallet
1 #2 rawhide mallet
1 Pair of tin shears
1 Knife
1 Flat file
1 Cold chisel
1 Lead scraper
1 Armature spoon
1 6 in. parallel plier
1 Set of soldering irons
1 Pair of scissors
1 Set of wedge drivers
1 Coil lifter and shaper
1 Long-nose plier
1 Diagonal plier
1 Set of coil tamping tools
1 8 in. side cutting plier
1 Hacksaw
1 Stripping tool for stripping open slot and stators
1 Coil hook to break coil ends loose from insulating varnish
1 Set of coil pullers for sliding top sides of coil into slots
1 Push cutter for trimming edges of slot insulation
1 Set of wire scrapers for removing insulation from ends of coil leads
1 Set of lead drifts for driving coil leads into commutator risers
1 Commutator pick for picking out short circuits between segments
1 Undercutting saw from undercutting commutator mica
1 Banding clamp for placing tension on banding wires while winding
   them

The use of most of the tools listed will be fully explained in the chapters on motor winding. See Fig. 18-1.

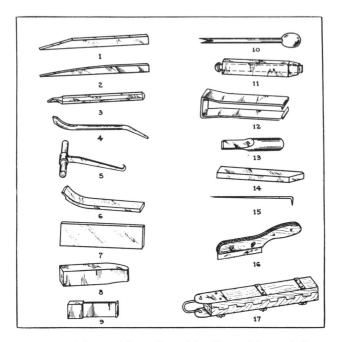

FIG. 18-1 Some of the specialty tools traditionally used for rewinding motors. (1) stripping tool for stripping open slot armatures and stators. (2) coil lifter for lifting coils from the slots. (3) lead lifter for lifting coil leads from commutator risers. (4) lifting tool for prying tight coils from slots. (5) coil hook to break coil ends loose from insulating varnish. (6) coil puller for sliding top sides of coils into slots. (7) fiber slot drift for driving coils into slots (four thicknesses needed: 3/16", 5/16", 7/16", and 9/16"). (8) fiber coil shaper for shaping coil ends after coils are in slots. (9) steel slot drift for driving coils to the bottom of partly closed slots. (10) push cutter for trimming edges of slot insulation. (11) wedge driver for driving wedges into partly closed slots. (12) wire scraper for removing insulation from ends of coil leads. (13) lead drift for driving coil leads into commutator risers. (14) one-sided chisel to cut off leads at risers. (15) commutator pick for picking out short circuits between segments. (16) under-cutting saw for under-cutting commutator mica. (17) banding clamp for placing tension on banding wires while winding them. (Courtesy Page Power Co.)

THUMM MACHINES

While the electric repair ·industry does not recognize the THUMM machine as the most economical or the most necessary equipment, Dreisilker Electric Motors, Inc. has been using this machine for their electric motor overhauling since 1968. It is used to strip electric motor stators and rotors as follows.

COIL CUTOFF AND COIL PULLING IS DONE ON ONE MA-CHINE. A stator is placed on the turntable of the THUMM machine with the connection side of the winding up. The stator is then quickly and firmly locked to the turntable. The cutoff motor with the saw blade is then lowered into the stator. The rotating saw blade is pressed into the winding just above the end of the laminations. Then by rotating the turntable the winding is quickly cut off. The stator is then taken off the turntable and placed on the stator warmer with the cutoff end down. The stator is heated up within minutes to soften and loosen the coils and insulation. The flame of the warmer will not touch the stator. The stator is then returned to the turntable and fastened down with the winding facing up. A hydraulic cylinder hooked to specially designed tongs will lift out all the coils and slot insulation with ease.

This process is one of the fastest, most economical ways to strip stators. No afterburners are required since no smoke is generated. No air purifiers are needed. The process leaves hardly any odor. Stators are not being distorted and can be wound immediately.

Armatures are stripped in the same fashion with the AZM-2000 rotor rebuilding machine.

The THUMM coil puller in Fig. 18-2 shows the unit in all its simplicity and ease of accessibility. The coil puller is available in four different models to accommodate stators from 1½ to 59 in. in diameter and rotors to 47 in. in diameter.

The THUMM coil puller with its self-contained hydraulic power pack system is the fastest and most modern coil cutoff and stripping machine in the world. All that the THUMM machine requires for operation is 230 or 460 V, three phase. It is the most versatile, compact, and rugged coil puller and stripper on the market today. This machine is an economical time- and labor-saving device.

TO CUT OFF THE WINDING HEAD. To cut off the winding head, the stator is placed on the ball-bearing-supported rotating platform. By turning the crank on the left side of the machine, the winding head cutter is positioned to the proper cutting height. The cutting wheel is then pressed back against the windings. By rotating the platform against the running direction of the cutting wheel, the winding head is cut. This procedure is repeated until the winding head is separated. The cutting wheel is then moved back to the center of the stator, raised, and swung out of the way.

WARMING OF STATORS. The warming of the stators can be done in a uniform, fast, and inexpensive way on the propane- or natural-gas-operated stator warmer.

LIFTING THE COILS. To lift the coils, the stator must be turned (coils up). With special lifting tongs the coils are firmly gripped and hy-

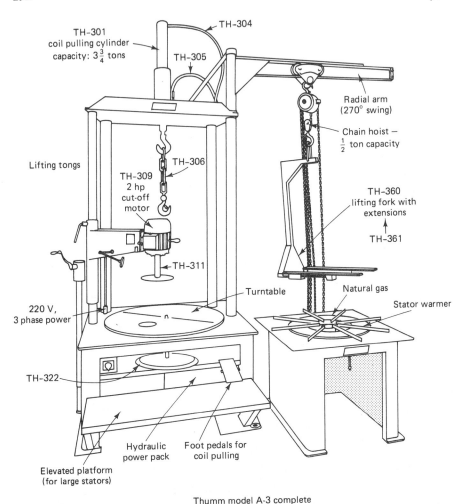

Thumm model A-3 complete

Capacity: $\frac{1}{10}$ to 350 hp

FIG. 18-2   THUMM Model A-3 complete. (Courtesy Dreisilker Electric Motors, Inc.)

draulically lifted so that complete coil groups are lifted out each time. The vertical movement is controlled by a foot lever, thus leaving both hands free to guide the lifting tongs.

The foregoing features are summarized and emphasized as follows:

1. Considerable savings in time and energy.
2. Coil cutoff and coil lifting are done on one machine.
3. Absolutely clean cutting and pulling.

4. Quick and secure locking of stators and rotors.

5. Vision is unrestricted since the fastening devices are under the table.

6. The platform is mounted on ball bearings; therefore, the stator is easily rotated.

7. Housing and lamination damage is entirely eliminated.

8. The gas stator warmer loosens coils and insulation without the use of chemicals.

9. Foot-lever-operated hydraulic system.

10. Compact and rugged construction.

The THUMM Model AZM-2000 rotor rebuilding machine (Fig. 18-3) is designed to strip large ac and dc rotors as well as large traction motor armatures. The machine comes complete with winding head cutoff and self-contained, hydraulic pulling cylinders. Standard or special tongs are available for pulling windings.

FIG. 18-3   The THUMM AZM-2000 rotor rebuilding machine is designed to strip large ac and dc rotors as well as large traction motor armatures. (Courtesy Dreisilker Electric Motors, Inc.)

The windings can be easily pulled out with either of two built-in hydraulic cylinders. The cylinders pull vertically and horizontally with 6615 lb (3000 kg) of pulling force and are operated by a foot pedal.

The rotor is placed between the chuck and tailstock. The variable-speed drive changes the rotating speed of the rotor to any required speed. The winding head cutoff unit is mounted in front of the headstock. It is adjustable and movable. With a wheel, it can be moved horizontally parallel to the rotor. By means of a handwheel, the position of the cutoff unit can be varied to fit rotors up to 47 in. (1200 mm) in diameter and 67 in. (1700 mm) long. The maximum rotor weight is 7700 lb (3500 kg).

The cutoff unit also can be rotated to bring the cutoff wheel parallel to the rotor (for undercutting commutators, etc.).

The rotor windings can be cut off on either side. Rotors can be warmed on the machine through an open gas flame or infrared heating.

Dimensions of the various models of THUMM machines are shown in Fig. 18-4.

Stripped frames by the THUMM method are shown in Fig. 18-5. Even aluminum frames, including new-design high-efficiency motors, can be stripped safely and easily with the THUMM method. No damage is done to either the aluminum frame or the laminations.

The THUMM model TA-1000-D/G machine (Fig. 18-6) allows a fast and efficient impregnation of stator windings with any solventless varnish. This design advantage eliminates time-consuming dipping and (oven) baking, thus facilitating more efficient motor repair.

The stator is mounted flat against the plate guided by the center bore, which guarantees an absolutely concentric position. The turntable with stator is automatically rotated and can be tilted to a 90° angle. The speed rotation is controlled by a single-speed, two-speed, or variable-speed motor.

The stator winding is heated through a three-phase variable-voltage transformer, and the trickling and/or hardening temperature is sensed and controlled without physical contact. Varnish is forced from the reservoir to the trickling jet by air pressure. All resins are usable with a viscosity up to 12,000 cps.

The advantages of this machine over the conventional methods are as follows:

1. Simple operation through automatic temperature and voltage control.

2. Leaves varnish only on the windings.

3. Continuous work procedure – stator can be processed immediately.

4. Reduced overall time provides faster deliveries.

5. No cleaning of metal parts necessary.

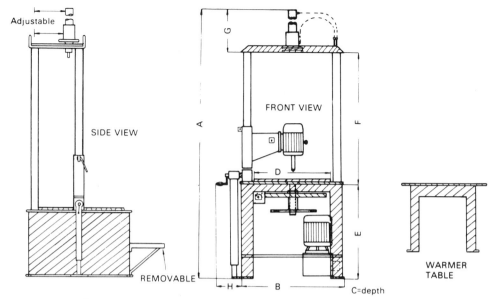

For your convenience dimensions are provided in inches and millimeters

MODEL	A	B	C	D	E	F	G	H
A2	90.55	31.50	27.56	23.62	29.53	43.31	20.47	7.87
A3	111.02	43.31	39.37	33.46	23.62	59.06	28.35	7.87
A5	137.80	85.43	66.93	59.06	23.62	70.87	43.31	7.87
A6	181.10	83.46	64.96	59.06	11.81	118.11	50.00	—

ABOVE DIMENSIONS ARE IN INCHES

MODEL	A	B	C	D	E	F	G	H
A2	2300	800	700	600	750	1100	520	200
A3	2820	1100	1000	850	600	1500	720	200
A5	3500	2170	1700	1500	600	1800	1100	200
A6	4600	2120	1650	1500	300	3000	1270	—

ABOVE DIMENSIONS ARE IN MILLIMETERS

FIG. 18-4 Dimensions of the various THUMM machines. (Courtesy Dreisilker Electric Motors, Inc.)

6. No resin loss.

7. Even and excellent slot fills.

8. Effective hardening even between fine wires and in long slots.

9. Total mechanical binding of windings.

10. High moisture resistance; no danger of causing water condensation in winding.

11. Better heat sink.

12. Varnishing of stators through the trickling method results in high quality and at the same time reduces cost.

13. Slip ring and dc armatures can be trickled.

FIG. 18-5 Motor frames stripped by the THUMM method. (Courtesy Dreisilker Electric Motors, Inc.)

FIG. 18-6 The THUMM Model TZ-1000-D/G machine allows a fast and efficient impregnation of stator windings with any solventless varnish. (Courtesy Dreisilker Electric Motors, Inc.)

ABISOFIX ISOLEX

The Abisofix (Fig. 18-7) is an all-purpose tool capable of removing the insulation on copper wire to facilitate terminating, that is, soldering the wire ends of stator and transformer coils and the like. This tool is simple to operate and easy to reset. The operator has only to select the correct speed step from a potentiometer for different wire sizes.

An electric motor drives an arrangement of three counterbalanced knives which close in around the wire by centrifugal force. The centrifugal force is governed by the speed of the motor, which in turn is controlled by a potentiometer. This speed control enables the tool to be set for various wire gauges; fine wires require low speed, whereas heavier wire requires higher speeds.

Because of the tool's compact design, it is possible to strip to within 1 mm of the soldering point. It is also possible to reach points on the terminals of motors and transformers which are not easily accessible.

The Abisofix can be used for removing the insulation from the following:

• Enameled copper wire

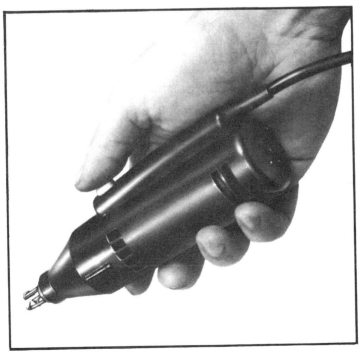

FIG. 18-7   The Abisofix is an all-purpose tool capable of removing the insulation on copper wire to facilitate terminating. (Courtesy Dreisilker Electric Motors, Inc.)

- Resistance wire
- Silk-, glass-, or polyester-covered wire – 35 to 11 AWG

With the addition of a special device, it can also be used for stranded wire (silk or plastic) from 30 to 10 AWG.

METAL TURNING LATHE

The metal turning lathe is a very valuable asset to any maintenance shop – so much so that no shop can really afford to be without one.

When selecting a lathe for motor repair, the most important factor to consider is the size of work that will be machined on the lathe. In general, this is determined by the greatest diameter and length of work. The lathe selected should have a swing capacity and distance between centers at least 10% greater than the largest job that will be handled. The buyer, however, must use discretion when purchasing a lathe for the machine shop. For example, if the majority of the armatures to be turned are less than, say, 7 in. in diameter – with only two or three motors in operation that are larger – it would be impractical to purchase a lathe to handle the larger motors which may only need overhauling once every 5 years or so. Since the majority of the work will be under 7 in. in diameter, a lathe with a 7-in. swing would suit the shop better. Then, when the larger motors required repairs, they could be sent to a large repair shop. By doing so, the maintenance shop could cut expenses considerably.

If the lathe required is a large one, say, with a 13-in. swing or more, the lathe should be equipped with its own supporting structure. On the other hand, if a 7- or 9-in. model is required, either a bench lathe or a small cabinet model may be selected.

For the repair shop doing work on small motor armatures or as a second lathe for the larger shops, no better lathe can be found for the price than the Myford Super 7 (Fig. 18-8). It will handle armatures up to 7 in. in diameter.

In overhauling armature commutators of electric motors, the commutator segments are first *trued* by taking a light, accurate cut with a carefully ground cutting tool. Then the mica insulation is undercut, usually by using a special motor-driven undercutting attachment designed especially for this application alone.

To perform this operation, the motor armature shafts – with center holes countersunk – are mounted between lathe centers and driven with the faceplate and lathe dog. In mounting the shaft, make certain that both center holes are free from burrs and dirt. If the shafts do not contain center holes, one end of the shaft is mounted in a three-jaw universal chuck, and the other end is supported by a Jacobs Center Rest Chuck. In

FIG. 18-8 The Myford Super 7 lathe is an excellent choice for the small shops or as a second lathe for the larger motor repair shops.

using the latter, tighten the bronze jaws just enough to remove looseness but not enough to cause *drag*. During the turning, be sure to apply plenty of lubricant on the shaft at the point of bearing with the jaws of the center rest chuck.

Besides being used for motor repair work, the lathe can also be used to perform many other useful shop tasks, such as turning and threading machine screws, making parts for motors or motor repair tools, and making custom tools to fit a particular need.

For further information on the Myford Super 7 lathe, contact D & M Model Engineering, P.O. Box 400, Western Springs, Illinois 60558.

MOTOR REPAIR MATERIALS

The proper insulation of a stator or armature means the insulation of the slots as well as the coils, the former serving the dual purpose of insulating and mechanically protecting the coils at the same time. These insulations may be divided into groups which indicate the purpose for which they are most suitable. In the first group may be listed strictly electrical insulations such as cotton tape, oiled cloth of cotton muslin or linen, varnished cambric, varnished muslin, varnished silk, and other types of insulating cloths.

In the second group, the materials affording the greatest mechanical protection include pressboard, hard fiber, vulcanized fiber, and fish paper. Those insulators especially adapted to high temperatures fall in the third group and include mica, micanite, mica paper, glass tape, and mica cloth. From this it would seem that there is an insulation for practically every purpose, and a certain degree of care must be exercised in choosing the insulation for any particular job. The best advice is to obtain the catalogs from the various suppliers of insulating materials and become completely familiar with each, reading the descriptions and specifications of each. Then the winder will have a good basic knowledge with which to make the proper selection.

19

Motor Repair

Before a motor fault can be determined, a test must be made to find out exactly what the problem is. The first step is to make sure the motor turns over easily by hand. You might be able to stop right here, as the trouble might be in the motor bearings, causing undue friction that will in turn cause the motor to heat up and, finally, burn out. Continue by checking for an open circuit, using a continuity tester; attach one of the test leads to the motor lead and the other test lead to another motor lead. If the indicator functions, the circuit is closed. Then touch one lead with one of the test leads and the other lead to the end of the motor shaft. If nothing occurs (no light, no needle swing, etc), the motor is not grounded. Should this test indicate a ground (the lamp of the tester lights, the needle swings, etc.), disassemble the motor and try to locate the fault.

Once the motor has been disassembled, always retest the leads at this time, as grounded leads are often the problem. If the winding is burned out, the burnt appearance may be seen easily when the motor is disassembled. The odor of burnt insulation is another sign that is easily detected. If the grounded part of the winding or leads cannot be found by this method and tests show that a ground does exist, the only alternative is to rewind the motor.

COLLECTING DATA

Assuming the armature is to be rewound, the next step is to collect sufficient data while dismantling the old winding. These data will enable the new winding to be installed correctly.

Begin by mounting the armature in suitable armature horses such as those shown in Fig. 19-1. Then, as the old winding is dismantled, mark the slots and commutator segments from which the first coil and leads are removed. This can be done with a conventional scratch awl or metal file. One small punch mark can be placed under the slot that held the top coil side and two dots under the slot that holds the bottom side of the same coil. The top leads are then traced out to the commutator, and each bar that they connect to should be marked with one dot. Next trace the bottom leads to the commutator, and each of the bars they connect to should be marked with two dots. This marking should be done to both lap and wave windings and is a positive way of keeping the core and the commutator marked so they may be reconnected properly. Keep in mind that the bottom lead will be the lead with which the coil is started and the top lead will be the ending lead once the coil is wound.

If deemed necessary, a sketch or diagram may be made of the first few coils removed to indicate which way the coils are wound. The sketch can be made similar to the ones in Fig. 19-2.

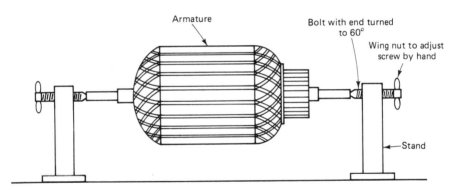

FIG. 19-1 When stripping and collecting data, armatures should be positioned in armature horses.

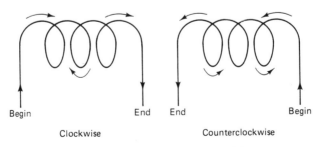

FIG. 19-2 A sketch or diagram of the first coil removed from a motor armature is one good way to record data from a motor.

In removing the old winding, remember that the coils will be compacted and not stretched out as shown in the diagrams in Fig. 19-2. This requires careful examination to determine just exactly how the coils are wound. For example, look at the two coils in Fig. 19-3. At first glance, they both look almost identical. However, a closer examination will reveal that coil A is wound right-handed while coil B is wound left-handed. When recording data for an armature, make certain that they are recorded exactly as found.

In addition to marking the core and commutator and keeping a diagram of the winding and connections, the following data should be carefully collected as the old winding is removed:

- Turns per element
- Size of conductor
- Insulation on conductor
- Coil insulation
- Slot insulation (layers, type, and thickness)
- Extension of slot insulation from each end of core
- Overall extension of the winding from the core, both front and back

If the preceding items are carefully observed and recorded, little difficulty should be encountered in replacing most windings correctly. It does, however, require practice to obtain coils of exactly the proper size and shape so that they will fit neatly and compactly in the armature. No book can do this for the winder. He or she must practice winding to obtain this ability.

Wire bands are sometimes used on large armatures having heavy coils to hold the coil ends securely in place. If the core has open slots, bands are often used over the core to hold the wedges in place. High-grade steel piano wire is commonly used for this purpose and can be obtained in various sizes from motor repair suppliers.

When a banding machine is not available, a lathe can be used to hold

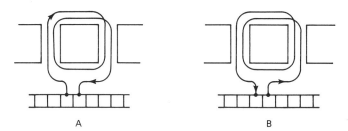

A   B

FIG. 19-3 Two coils that look identical may in all probability be entirely different. Check the coils thoroughly and be sure of your findings before recording data.

the armature while the bands are wound on. A layer of paper or cloth is usually placed under the band. Cloth makes the best foundation for bands placed on the coils ends, as the cloth tends to keep the bands from slipping off. A layer of fish paper can be used under bands placed around the core. Grooves about $\frac{1}{32}$ in. deep are usually provided for the bands on cores with open slots.

The paper should be cut carefully to the exact width of this groove, so it will fit snugly and without sticking out at either side. The banding wires should be wound on under tension, so they will be firm and tight when completed. A simple tension clamp or brake can be made by cutting two strips of fiber $\frac{1}{4}$ by $1\frac{1}{2}$ by 6 in. and bolting them together with two small bolts, using wing nuts on each end. Place these pieces of fiber in the tool post of the lathe and run the wire between them at very slow speed. Then, by adjusting the two wing nuts, any desired tension may be obtained.

To start the first band, make a hook of heavier wire and attach the band wire securely to this hook. Then slip the hook under the ends of a couple of coils close to the ends of the slots and start winding the band wire on the core. Make two or three gradual turns around the core to get the band wire over to the first slot. As the first turn is wound in the slot, narrow strips of tin should be placed in the slot under it and every few inches apart around the core. Drawing the first turn tight will hold these strips in place, and other turns are then wound on over them. Wire should be wound with the turns tightly together until this groove is full. Then fold up the ends of several of the tin strips to hold these wires in place, run the wire across to the next groove with a couple of gradual turns around the core, and start the next band without cutting the wire. Continue in this manner until all the bands are on. Then, before releasing the tension on the wire, run a thin layer of solder across each group of band wires in several places to keep them from loosening when the end wires are cut.

After cutting the wires between the bands, cut these ends off to the proper length, so that they will come directly under one of the tin clamping strips. Then fold in the ends of all these strips tightly and solder them down with a thin layer of solder.

These tin strips are usually about 15 mils thick and $\frac{1}{4}$ in. wide and should be cut just long enough so that their ends will fold back over the bands about $\frac{1}{4}$ in.

While stripping armatures, note that one coil is on top of the others in the usual arrangement. Unwinding the top coils would reveal another coil which is now on top of the remaining coils and so on until the last coil is reached. Such coil progressions are known as either right or left.

For example, assume that the first coil is placed in a slot, that the next coil is placed in the slot to the right of the first one, that the next coil is placed to the right of that one, and so on. This progression is wound to the right and is known by the same name. In winding a motor to the right, it is necessary to turn the armature toward the left after each coil is wound into the slots. In all cases, the armature should be positioned so that the commutator faces the winder – in either stripping or winding – and all measurements and figures are taken with the armature in this position.

If it is desired to have the coils progressing to the left, it becomes necessary to turn the armature to the right after each coil is placed in the slots. The first coil is placed in its slots with each successive coil being wound into the slots toward the left.

The direction in which coils progress has no influence on the direction of rotation. The choice is a matter of preference depending on which way is easier for the winder.

LEAD CONNECTIONS

In collecting data from the old windings, make sure that the lead connections are accurately recorded. This can be accomplished by using some or all of the following suggestions:

1. Find the top side of the coil. If the armature is wound with one coil upon another, take the side of the coil which has the ending lead. If the armature is of the form type, more care must be taken in determining the lead connections as one side of each coil is in the top of the slot and one side is in the bottom. In the latter type, the top leads are usually separated from the bottom ones by tape. Ascertain the number of coils per slot and remove that number of leads from the segments. The coils per slot may be calculated by the following equation:

$$\text{coils per slot} = \frac{\text{number of commutator segments}}{\text{number of slots}}$$

2. Mark the segment which contains the leads that have been removed. Make a loop in each of the leads and leave that coil momentarily. Cut the back leads of enough coils so that they may be pulled out until the coil with the looped leads is whole. Be sure not to disturb the bottom leads on removing the coils. Unwind the coil until the other leads of the coil come to the commutator and record where they connect to the bars.

3. Remove the top leads as directed and make loops in them. Desolder and pull out all top leads. Use a test lamp to touch each of the looped leads with one test lead and the commutator with the

other lead. When the light lights, that is the commutator bar to which the bottom lead connects. Check carefully to make certain that the test lamp does not light on another bar also. If so, clean out mica between bars and test again.

4.   Mark the bars where each of the top and bottom leads for the number of coils per slot connect. Remove enough coils to remove the coil with the looped leads as a whole coil. If these loops are on the ending side or top of the coil, then these leads with the loops are sure to be the top leads.

5.   In some rare cases, the bottom leads may be found on top of the winding. Therefore, the coil should be unwound to be certain where the beginning and end of the leads are located. Remember that beginning leads are bottom leads and that ending leads are top leads; record them as such.

Determining lead connections for conventional windings is a much easier task. Merely follow the lead to the commutator and mark the top lead; the number will depend on the number of coils per slot. Next unwind the coil until the bottom leads are found. Mark the commutator where they are located with a light punch or scratch awl, making certain they will not be erased easily. Use one mark for the top leads and two marks for the bottom leads as discussed previously.

Some winders prefer to use a test lamp to determine the various leads, but this method is not always reliable. Some armature windings can be so badly burned that a test lamp will not give accurate readings. Therefore, it is always best to unwind a coil to collect the data.

RECORDING DATA

Once the leads have been located, take the slot having the top leads and — with a piece of metal — lay it in the center of the slot so it will extend the length of the slot and to the commutator. A straightedge such as a metal rule is ideal for this step. Just be sure to position it in the center of the slot. The bar on which the straightedge strikes the commutator is known as bar 1 and the next bar as 2. If the top lead falls several bars either to the right or left of the bar straight out from the slot, the top lead connection will be so many to the right or left, counting this bar as 1, the next as 2, and so on.

When there is more than one coil per slot, follow the same procedure, except list the data with the top lead nearest bar 1 first.

After finding the top lead connection, locate the bottom lead. In a lap-wound armature, where the bottom leads connect out next to the top leads, record "bottom leads next to top leads" on the left or right depending on which side of the tops they are, or record "bottom leads adjacent to top leads" on the right or left.

The following suggestions may help in recording these data:

1. Locate the bars in which the bottom leads connect, and beginning with bar 1 (and counting it as 1), count over as was done for the top leads – only, however, to where the bottom leads connect.

2. With a wave winding, the top leads and bottom leads are connected to commutator bars at quite a number of bars away from each other. Collecting data for this type of connection varies somewhat from that for a lap winding, and there are two methods normally used to record these data:

a. After having found the top lead connections, count from the bar having the top lead and designate this bar as bar 1 until encountering the bar that has the bottom lead for that coil. This is known as the commutator span, between top and bottom leads, and the bars containing these leads are counted when arriving at this commutator span.

b. Take the bottom side of the coil or the side where the bottom leads of the coil enter and place a straightedge in the center of the slot out to the commutator, as done previously. As soon as bar 1 is located, count over toward the direction of the bottom leads. The bottom leads may be 5 and 6 left. The data would then read "tops 1 and 2 right, bottoms 5 and 6 left, wave wound."

Diagrams such as Fig. 19-4 help to show more clearly the various connections and how they are made. Diagrams also eliminate the need for a lot of written instructions, saving the winder much time.

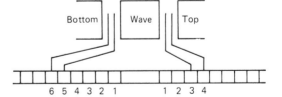

FIG. 19-4 A sketch such as this one can save much time in recording coil connections.

DATA CARDS

Recording data from a motor nameplate and/or by tests and examinations has been successfully accomplished in several different ways. Some repairmen merely write out the data on a card and possibly accompany these data with a wiring diagram or two. A sample of these data follows:

General Electric ½ hp. Type RSA, frame 415, form D, 120/240 V, 10/5 A, 1750 rpm, one phase, 60 cycles, serial number 76894. Armature, 21 slots, 41 bars, 43 coils, with 1 dead coil. Wound right-handed, coils progress right. Top lead 4, 5 right, bottoms 5, 6 left, commutator span 1–21. 12 turns #18 SOE, coil span 1–7. Retrogressive connections.

In shops doing much motor repair work, however, a standard form is better. In general, such a form is a systematic method of listing information concerning motors that are repaired. When properly organized and thoroughly understood, these forms are not only time-saving devices but can help prevent an oversight in collecting data, since there is a section for all necessary information. The information on the card will be the same as described in the preceding example.

20

Motor Repair Techniques

Before a repair can be made to an electric motor, the fault or faults must first be determined and then a logical plan devised to accomplish the required work. Although the exact procedure for repairing motors will vary from motor to motor (and from fault to fault), the following is a very general outline of what a motor repair job entails:

- Make all necessary tests and disassemble motor.
- Strip the armature and collect data, making a data card during the process.
- Clean and test the commutator, installng a new one if required.
- Remove old insulation from the armature.
- Wind the coils on the armature core.
- Connect and solder leads.
- Dip and bake the armature.
- Turn down the commutator and balance the armature.
- Assemble and test.

You might say that this description is a complete overhaul of a motor. Some repairs, however, will require only the replacement of, say, a commutator, a set of brushes, or the like, and obviously the entire process (as outlined) will not be necessary.

Unless an armature winding is definitely known to be defective (beyond repair) or burned out, it is advisable to test it before tearing it out. Sometimes while testing, defects may show up, but a closer examina-

209

tion will show that the problem is due to some foreign material that has found its way into the machine – such as a piece of copper between the bars of the commutator. By cleaning this material from within the motor, a second test may show everything to be in working order.

There are numerous tests for motor armatures, most of which have been covered in previous chapters.

COMMUTATORS

Commutators on many motors will be found to be in very bad condition and must be replaced. but before removing the old cummutator, measurements must be taken to ensure that the new commutator will be in the correct position for the motor to operate properly. The main measurement is the distance from the end of the shaft to the front edge of the commutator as shown in Fig. 20-1.

If the commutator is of the vertical type, allowance should be made for wear in the old commutator.

To remove the commutator from the armature shaft without damaging the shaft or stator laminations, first examine the commutator for fasteners such as pins, screws, or other devices which are holding it to the shaft. Remove any found.

Lay two bars in back of the commutator, and with an arbor press, apply pressure to the shaft and push the shaft through the commutator. Always put a soft piece of metal on the end of the shaft so it will not be damaged. If no press is available, a hammer may be used to pound the commutator off, but be extremely careful not to chip the shaft by pushing too hard on one side. The armature should be turned as the pounding is done. Of course, if you use the pounding method, the commutator will be damaged beyond repair and cannot be used again; a new one will have to be installed. In some cases, pulley pullers can be used to good advantage in removing old commutators from armatures shafts.

To install a new commutator, reverse the process. Of course, more care must be used in this installation because it must not be damaged at all. The best method is to press the new commutator on with an arbor or hydraulic press, but many have been installed with a length of pipe. The inside of the pipe should be just large enough to allow the shaft to enter but not so large as to rest on the bars of the commutator.

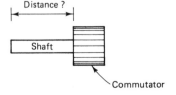

FIG. 20-1 Before removing the commutator from a motor shaft, take measurements to ensure that the new commutator will be installed correctly.

Another method is to start driving the commutator on lightly with a hammer and then turn the armature over on a place which has a hole in it for the shaft to go through. Press on the other end of the shaft if it will stand pressing and not spring. If it looks as if it would spring, then press against the armature laminations. If no press is available, a wood block held against the commutator which in turn is lightly tapped with a hammer will work most of the time. Regardless of the method used, be extremely careful with this phase of the repair.

Once the armature has been stripped as described in Chapter 18 and the existing commutator is to be used, clean out the mica between the commutator bars by scratching it lightly with a knife blade file or a hacksaw blade. Then use a tester between the commutator bars. If a reading is obtained, clean the slots more as there are probably metal filings present or carbon deposits or excess oil.

An extremely dirty commutator should be degreased with AWA 1,1,1 and then blown dry with compressed air before any tests are made. Of course, the wires should also be removed before making the tests.

The slots for the new windings may be cleaned out with a hacksaw or small file, or a special cutter is available from motor repair supply houses with a sharp point designed to chisel out old wires and solder. When the slots are clean, a medium-sized flat mill file is used to file over the top of the slots to remove older solder there, making it easier for the new solder to "take" when the new leads are connected to the bars.

REMOVING INSULATION

Old insulation must be removed to make room for the new and also to get rid of any that is damaged so as not to short out the new windings. The removal is easily done with a hacksaw blade beveled at its width. It is used in the slot much like a shovel is used to straighten a hole in the earth. The diagrams in Fig. 20-2 show how this is accomplished. A pocketknife may also be used to scrape down sides as shown in Fig. 20-3.

In removing the insulation, note if the insulation around the shaft and at the end of the core is in good shape. If so, do not disturb it; just coat it with schellac to preserve it better. If it is burned or otherwise

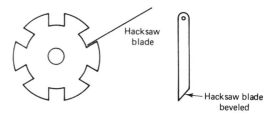

FIG. 20-2 Old insulation may be removed with a beveled hacksaw blade used in a way similar to a shovel straightening a hole in the ground.

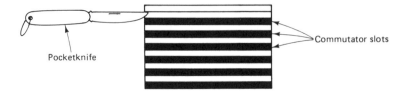

FIG. 20-3   A pocketknife may also be used to scrape down the sides of commutator slots.

defective, it will naturally have to be removed along with the other insulation.

After all slots have been cleaned, line up laminations properly so
they will not cut the new winding when it is installed. File down any
round spots on the laminated parts around the slots. Abrasive paper may
do the job if the area is not too rough. Laminations are the small sheets of
steel that make up the stator.

COIL AND SLOT INSULATION

In addition to the insulation on the wires themselves, it is also necessary
to insulate the coils and entire winding from the slots and armature core.
The insulation used for this purpose serves both to protect the coils from
mechanical injury from contact with slot edges and also to electrically insulate them from the slots. The materials commonly used for mechanical
protection include fiberglass, fish paper, manila paper, fiberboard, etc.

For heat-resisting and electrical insulation, mica, micanite, mica
paper, and mica cloth have been used quite extensively. Mica is a mineral
which is mined in flake or sheet form and is one of the very few materials
which will maintain a high dielectric strength at high temperatures. However, it is not very strong mechanically in its original form, but it is generally made up in sheets by cementing numerous thin flakes together.
This is commonly called *micanite* and is used for insulating armature
slots, between high-voltage coils, and for commutator insulation. Flexible sheets are made by cementing mica splittings or flakes to paper or
cloth.

When reinsulating the armature exclusive of the slots, there must be
enough thickness in the insulation around the shaft and on the face of the
laminations to prevent a ground.

After insulating discs have been cut and placed in back of the commutator, several thicknesses of friction tape should be used around the
shaft and also where the laminations show to compensate for the destroyed end fibers. This tape also helps hold the insulating discs in place
and insulates the winding from the shaft. Once in place, it should be

coated with shellac to secure it as well as to add to the insulating qualities.

Before any actual winding takes place, slot papers must be cut to size – the ones that protect the coil from becoming grounded to the laminated core. These papers must be thick enough to withstand tearing caused by vibration, and they must not be so thick that the coil will not enter into the slot. Many winders like to save a sample of the old slot insulation to use as a guide when matching the new insulation.

Most armatures use one thickness of insulation paper of the fiber type and one piece of oiled linen or varnished cambric. On 120-V armatures, the winding can function satisfactorily with one .007-in. paper and one .007-in. piece of linen. If there is sufficient room, one .010-in. paper and one .007-in. piece of linen may be used. For 240 V, at least one .010-in. paper and one .010-in. piece of linen must be used. Some armatures require the use of .015-in. insulation papers, while others use two papers and one piece of linen.

When two papers and one piece of linen are used, one of the papers is cut the same size as the linen cloth. One paper is always used as a feeder paper, since it sticks above the slot so the wire may be fed into the slot without injuring it on the laminations.

The linen and regular papers are cut so the width is such that the paper is approximately flush with the top of the slot, but an ⅛ in. or so of protrusion won't hurt anything; in fact, most protrude about this distance.

The feeder paper is the same length as those of the linen and regular papers, but its width is somewhat different, since it serves as a feeder paper also. The width of the feeder is cut so that it protrudes about ½ in. above the slot.

The insulation paper is very important to the life of a motor, and a good grade of paper should always be used. Standard slot insulating paper is tough, absorbs varnish well, and is practically moisture free. It also has strong dielectric qualities not present in wrapping paper and cardboard.

Some motor repair shops stick to the practice of insulating between coils, using a separator which is nothing more than a piece of regular fiber-type insulation paper cut slightly wider than the slot and about ⅛ in. longer. This paper is then bent in the middle to form a V the length of the slot. As soon as the coil is in the bottom part of the slot, the inverted V is placed in the slot and tamped down, forming a seal between the top coil half and the bottom coil half. The diagram in Fig. 20-4 illustrates how these coil separators are installed.

Another place to insulate between coils is where they intersect on the ends outside the laminated core part. This is done by laying a piece of oiled linen or cotton tape across the side of the coil just wound, so that

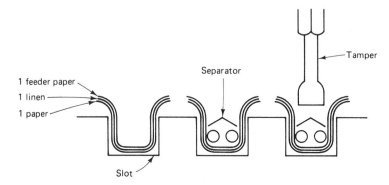

FIG. 20-4   Method of insulating slots.

when the next coil is wound, it may lay on the top of tape instead of the coil below it. Nearly all plain enamel wire-wound armatures, as well as many others, use this method between coils.

Insulating between coils is compulsory on some jobs, while the practice is optional on others. Inspect the armature carefully to see what procedures are required before it is stripped. After doing so, if none of the preceding insulating techniques are necessary, it will be at the winder's discretion whether they should be used or not.

Some armatures require that the leads be sleeved; others do not, and still some others leave sleeving up to the winder. There are many ways of sleeving the leads, such as just sleeving the bottoms and not the tops and using different colors of sleeves for each of the coils where there is more than one coil per slot.

Put the sleeve on the bottom, or beginning lead, and wind the coil. When the coil is completed, put the sleeve on the ending, or top lead. Each coil is handled the same way, making the sleeves long enough to extend back into the slot, so they cannot be pulled off easily. It is alright if the sleeves are long enough to extend beyond the commutator; they may be cut back even with the back of the commutator.

Where there is only one sleeve used, it is usually put on the beginning or bottom lead as the leads lay next to each other and may be shorted easily. The bottom leads lie on the top of this cloth separator. Since the top leads are easily seen, they may be kept separated from each other and therefore need no sleeving on them. Bring them out as straight as possible from the slot to the bar, making sure they are separated from each other.

Where more than one coil per slot is present, each sleeve should be given a different color, and the sequence of this color coding should be kept the same for all slots, so leads may be laid easily with no confusion. For example, the leads of the first coil could be left uncolored, while the leads of the second coil could be colored, say, red; the third, blue; and so on. These leads can be colored as they are installed on the wire during winding.

Different types of armatures will present different problems. For example, some are wound by laying the bottom lead back and then bringing it through the same slot as the top lead is brought out. This is accomplished by starting the winding from the back of the armature or the side opposite the commutator, beginning with the top slot instead of the bottom. The lead is left long enough to be brought back through the slot in the commutator. Such leads should be sleeved in such a way as to go from the back of the armature directly to the commutator. In winding this type of armature, be certain to arrange the top and bottom leads so they may easily be found. String markers and color coding are the two most often used ways of identifying them.

Some armatures are wound by starting out with a bottom (beginning) lead. Instead of cutting off the wire when the top lead is reached, the winder carries it back to the next slot where the next coil will begin. The wire is held in place with a finger while it is drawn out to reach the commutator, where it is then twisted around to hold it in place. This will result in the top lead of the first coil and the bottom lead of the next coil being one lead. Sleeve the leads as they are pulled and twist them so when they are sleeved they will lay back in the slot somewhat as shown in Fig. 20-5. This will help to hold the sleeving from the lead in the slot until the next coil can be wound over it.

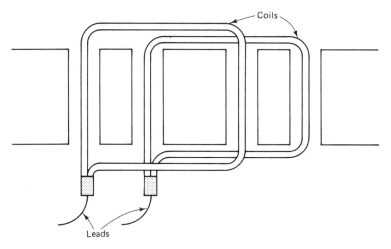

FIG. 20-5   Twist the leads so when they are sleeved they will lay back into the slot.

ac Motor Windings

SPIRAL-TYPE, LAP, AND WAVE WINDINGS

Three of the commonly used types of windings for ac motors are the spiral-type, lap, and wave windings. The spiral type has been used rather extensively on small single-phase motors and gets its name from the poles being wound in a spiral form as shown in Fig. 21-1.

In winding a spiral-type stator, the wire is started in the two slots to be used as the center of a pole, and after winding the desired number of turns in this coil, the process is continued on in the same direction in the next pair of slots with the same wire. In this manner, the coil for one pole is built up, working from the center to the outside. Sometimes more than one slot is left empty in the center as the first winding is inserted.

Another method used in the past is known as skein winding, shown in Fig. 21-2. In this method, the long skein coil is first made up of the right number of turns and the proper length to form the several coils. The end of this skein is then laid in the center slots as shown at A in Fig. 21-2, and the long end given one-half twist near the ends of the slots, as shown at B. The remaining end is then laid back through the next two slots — at C — and again twisted one-half turn so its sides cross near the first coil end. Then the last loop is laid back through the outer two slots to complete the coils for this pole.

The skein method of winding is quite a time saver where a number of stators of the same size and type are to be wound. After carefully measuring to get the first skein coil the right length, the balance of the coils can be made on the same form and the stator poles wound very rapidly.

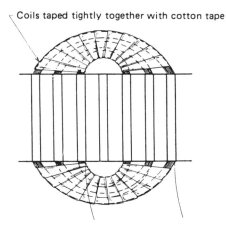

Coils taped tightly together with cotton tape

FIG. 21-1   Winding for a spiral-type stator. (Courtesy Page Power Co.)

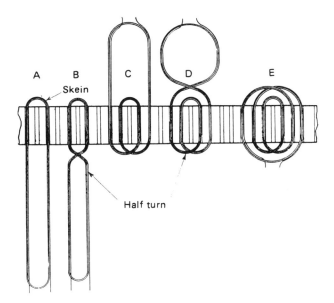

FIG. 21-2   Skein windings are sometimes used when winding a number of stators exactly alike. (Courtesy Page Power Co.)

Both lap and wave windings are also used for ac motors and alternators, but some of the rules given in Chapter 7 for dc motor windings do not apply here. For example, instead of classifying them as parallel and series windings, as done for dc windings, they are defined for ac motors as follows:

- A *lap winding* is one in which all coils in a pole group can be traced through before leaving that group.

- A *wave winding* is one in which only one coil in each pole group can be traced through before leaving the group.

When used in dc motors, the wave connection gives the highest voltage. This is not true of ac motors, as the wave connection gives no higher voltage than the lap. A single-circuit ac lap winding puts all possible coils in series, so it gives just as high a voltage as the wave winding. The wave winding, however, is considered stronger mechanically than the lap winding and for this reason is generally used for phase-wound rotors, as there is often considerable stress on their windings due to centrifugal force and starting torque.

Stators are usually wound with lap windings. However, in ac stators, the number of slots is determined by their size and the number of poles and is selected for convenience in connecting the type of winding desired for the purpose of the machine.

COIL POLARITY

When there is more than one coil per group in a motor winding, care must be given to the coil's connections, as all coils of the same group must be connected for the same polarity – in other words, so the current flows in the same direction through all coils of the group.

Figure 21-3 shows a simple two-pole, three-phase winding with one coil per phase group and three groups per pole. This winding only has one coil per group. Note that the groups are connected to give alternate polarity, that is, north, south, north, etc. Also note that two coil sides per slot are present, one lying on top of the other.

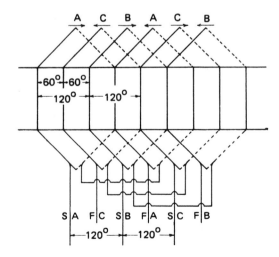

FIG. 21-3 Note the spacing in degrees between the coil sides in this two-pole, three-phase winding. The arrangement of the coil connections is also important. (Courtesy Page Power Co.)

The leads from the coil ends are referred to as top and bottom leads, and as the names imply, the lead from a coil side lying in the top of the slot is called the *top lead*, and the one from the bottom coil side is called the *bottom lead*.

In making the connections from one pole group to the next of the same phase, always connect like leads together, that is, bottom leads together and top leads together for the short jumper arrangement. This method should be followed to the letter to produce the alternate poles which are necessary in the winding to make the machine operate. If any of these coils are connected incorrectly, the coils will overheat, as their self-induction will be neutralized and too much current will flow.

TYPES OF COILS

Stators of 15 hp and under, and for less than 600 V, usually have partly closed slots and are commonly wound directly into the motor by the *fed-in* method. Two types of coils are normally employed in this method: the threaded-in diamond and the basket coil, also known by other names in different shops.

Figure 21-4 shows these two types of coils which are commonly used in winding small stators with partly closed slots, as these coils can be easily fed into the narrow slot openings.

The diamond coil is wound and shaped and the ends taped with half-lapped cotton tape before the coil is fed in the slots. The basket coil is simply wound to the approximate shape, and to the proper length and size, but is left untaped except for little strips of tape at the corners to hold the wires together until they are placed in the slots. The ends of these coils are taped after they are placed in the slots, or in some cases on small stators, the coil ends are left untaped. After placing the coils in the slots, their ends are shaped with a fiber drift and a rubber or rawhide mallet so the coil ends can pass over each other.

The basket coils described in this section are better suited for the smallest motors, while the fed-in coils are more suited to the larger machines under, say, 15 hp. The untaped sides of either of these types of

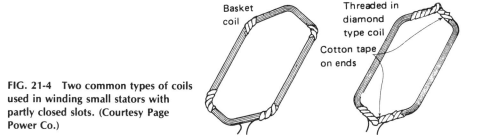

FIG. 21-4 Two common types of coils used in winding small stators with partly closed slots. (Courtesy Page Power Co.)

coils make it possible to feed the wires one or two at a time into the narrow slot openings.

WINDING FAULTS

The greatest number of defects occurring in motor windings during service or operation are caused by short circuits, open circuits, and ground faults. Water sometimes finds its way into the coils, or oil from the bearings may have destroyed the quality of the insulation. Metallic dust and grit sometimes work into the windings and cause short circuits, or a static charge from lightning or other sources may cause punctures or small pin holes in the insulation, which result in flashovers and ground faults.

Any one of the faults mentioned in the previous paragraph is also likely to show up just after a motor has been repaired or overhauled. Therefore, all machines should be carefully checked before putting them back into service. If a motor does not operate properly after being overhauled, it is quite likely that some of the coils are connected incorrectly or that there is a short, open, or ground in some coils because of careless work during the repair.

When running, most small induction motors are almost noiseless, and even in the larger machines, only a uniform, gentle humming should be heard. This humming noise is due mostly to vibrations of core laminations, which are caused to vibrate slightly by the reversals of the magnetic field. In addition to this humming, which is unavoidable even in motors in perfect condition, there might also be a slight whistling sound caused by the fan blades on the rotor, friction of the air with the revolving parts, and air passing through ventilation ducts. This air whistling is harmless and will continue for a short period after the motor is deenergized and still turning. If a motor is unusually noisy, there is probably some defect responsible for the noise.

A deep, heavy growling is usually caused by some electrical trouble, resulting in an unbalanced condition of the magnetic field in the windings.

If a shock is felt upon touching the motor frame, it is almost certain that one or more of the coils in the winding are grounded to the core or frame. This fault can be extremely dangerous, and for this reason it is very important that all motor frames be grounded when the machine is installed. If a properly grounded coil does become shorted or grounded with the frame, the overcurrent protection normally activates at once, protecting both the motor and persons who may be in contact with the frame of the motor.

22

Motor Winding

Motor coils may be wound with a single wire or several wires in hand, but in either case, the coils are wound practically the same except that more forming is required where there are several wires in hand.

Sometimes it is desirable to use two or more wires in hand instead of a single wire because the smaller wires are easier to form than the larger single wire. Also, the shop may not have the proper size larger wire in stock, whereas the smaller multiwires can be substituted. The substitution is made by dropping the size to one-third size below the larger wire. For example, if one AWG 16 wire is required, two 18 AWG wires may be used instead.

Some winders wind the coils with two or three wires in hand where there are two or more coils per slot. This saves time because the winder can wind more than one coil at a time. In using this method, sleeve the wire with both the beginning (bottom) lead and the ending (top) lead sleeves color-coded for the separate coils. Leave the sleeves for the beginning leads, and slide the other sleeves along the wire until the required number of turns for the coil is reached. Leave the sleeves as the top lead sleeves.

If it is not desirable to carry the sleeves on the wire, merely color the leads as the winding progresses, using a test lamp to find the top lead and the corresponding bottom lead.

The length of the coil is sometimes determined by winding a sample coil to obtain the proper length. The wires for the sides are sleeved with the proper colored sleeves, or colored the proper colors. The coils are then wound with precut wires, saving time in sleeving and making all the coils the same length. It also eliminates the necessity of dragging the wire from the spool as the motor is being wound.

Before starting the procedure of wiring or winding ac motors, the following terms and definitions should be thoroughly understood.

Coil group: The number of coils for one phase for one pole. The equation for determining a coil group is therefore

$$\text{coils per group} = \frac{\text{slots}}{\text{poles} \times \text{phases}}$$

Full-pitch coil span: The coils that span from a slot in one pole to a corresponding slot or position in the next pole. The equation for determining the full-pitch coil span is

$$\text{full-pitch coil span} = \frac{\text{slots}}{\text{poles}} + 1$$

In general, full pitch is 100% pitch. In some cases, however, a winding may be more than full pitch but seldom, if ever, exceeds 150% pitch.

Electrical degrees per slot: The term commonly used to express the portion of the pole which one slot covers. The equation for determining the electrical degrees per slot is

$$\text{electrical degrees per slot} = \frac{180 \times \text{poles}}{\text{number of slots}}$$

MOTOR WINDING

Once the slots have been insulated as discussed in Chapter 20, begin the winding by placing the armature in a rack or other suitable mounting device so that the commutator faces the winder. The wire is then taken in either hand – whichever is more convenient – as it is wound from a suitable rack or in its proper length as in the case of precut coils. A piece of sleeving is placed over the wire, and the end of the wire is placed so it faces the commutator.

The winder continues by winding the wire in the slot, beginning with the wire entering between the slot papers of the slot containing the bottom side of the coil and continuing on through that slot to the back end of the armature (the end opposite the commutator). The wire is then brought across the back of the armature to the other slot containing the other side of the coil span. It enters this slot from the back and is run through to the commutator end of the armature, where it crosses and re-enters the slot which it previously passed through. This procedure is repeated until the proper number of turns per coil is obtained, after which

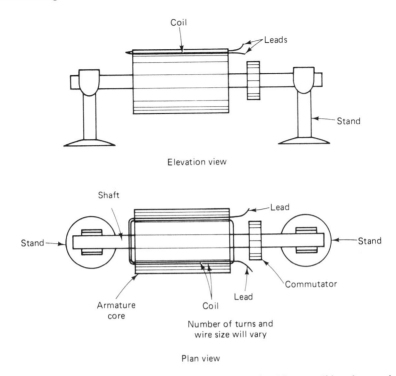

Elevation view

Plan view

FIG. 22-1   Motor armature placed in armature stands with one coil hand-wound
in place.

the wire is brought out to the commutator and cut off. A sleeve placed on
this end of the wire completes the coil. See Fig. 22-1.

Although Fig. 22-1 shows only 2 strands of wire, in actual practice
many more will be present, such as 15 strands or turns. Once one coil is
wound in the manner already described, the leads are cut to length so
they will reach the commutator, and then the leads are sleeved. This same
procedure is repeated until the required number of coils have been wound,
filling every slot.

To keep the windings looking neat, the first few coils should be
wound so they will build out from the armature core toward the end of the
shaft as shown in Fig. 22-2 and wound as close to the shaft as possible.
After the third coil has been completed, the remaining coils should be
wound so they will build up from the shaft toward the circumference instead of out toward the end of the shaft. However, never allow the windings to build up or out too much as this will allow them to be out of balance and become unsightly. Always wind the coils so as to keep the end of
the winding, as well as the roundness, uniform. This can be accomplished
by spreading or compacting the coils as they are wound, whichever way is
necessary to keep them uniform.

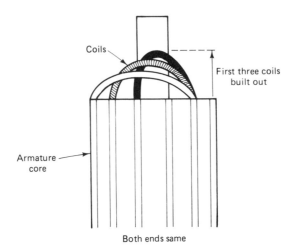

Both ends same

FIG. 22-2   The first three coils wound on a motor armature should be spaced at different distances from the armature core.

Besides being uniform and neat, the winding must be firm at the ends of the slots where the coil has to travel across for the other coil side. This can be done by pressing each turn against the end of the winding with the thumb as it is wound.

When winding the coils in the slots, make certain that the slot papers do not slide, causing the winding to touch the stator laminations on one side; if so, they are likely to become shorted or grounded. Just keep an eye on the papers as the work progresses and adjust any that slip. Also try to keep all the papers in the slots so the end of the paper is the same distance from one lamination on the coil as it is from another.

One of the best ways to understand how armatures are wound is to watch an experienced winder perform the various steps. If this is not possible, examine some armatures that are already wound. Before removing old windings from an armature, carefully study how they are placed into the various slots and then make notes that can be referred to as the new windings are installed.

Before the winding is completed, slot sticks must be installed in the slots to hold the coils in place. These sticks can be put in as the armature is being wound or when the winding job is completed. In cases where the armature leads are brought back through the same slots (top and bottom leads in same slot), it is best to wait until the winding is completed before the sticks are installed. In conventional windings, however, the sticks may be installed at any time after any slot has both coil sides in it for the proper number of coils per slot. In either case, cut off the feeder paper even with the top of the slot, being sure the wires stick up so they may be cut into when the paper is cut off. Fold each side of the paper over the winding to form an envelope, which holds the wires inside the paper.

Push or lightly tap a slot stick in over the folded papers, holding them securely down. If the slot stick goes in loosely, two may be required to adequately hold the coils in place. If two cannot be placed in the slot, take part of the paper that was previously cut off and place it in the slot full length; this will tighten the stick. The stick itself should be the same length as the paper.

Anything that will hold the coil in the slot is permissible, but manufactured slot sticks just for the purpose are better as they are easier to use and will usually offer a better, more uniform fit.

FORM-WOUND ARMATURES

Since hand-wound armatures are not gauged, so to speak, coils will naturally not be of the same size when the job is completed. For this reason, the hand-wound armatures may be slightly out of balance, and often weights must be added to balance them. Form-wound armatures, on the other hand, are rarely out of balance due to the method by which they are wound and installed.

In form winding, the diamond coil is probably the most frequently used, but other shapes can be used if necessary. The type of coil selected and the way it is wound will depend largely on the type of old coil removed from the motor, so be sure to save one of the existing coils as a guide during the stripping of the old winding. In case the coil in the armature is of the pulled type with a twisted knuckle, save at least two samples of the old winding, one showing the coil when twisted and the other straightened out to act as a template for making a coil form.

The leads in a form-wound coil may come out any number of places. Regardless of how the coil may be wound and taped or how the leads come out and other factors, the new coil should be an exact duplicate of the old ones.

The equipment for making diamond-shaped coils with twisted ends include a loop coil winder, some tin or lead cleats, pattern pliers, and two hardwood blocks with grooves about ¾ in. deep and the width of the coil. The length of these blocks should be approximately ½–¾ in. longer than the length of slots through the armature.

Where there is more than one coil per slot, the amount of wire in hand is the same as the number of coils per slot. These coils are wound on the loop winder tightly and uniformly (see Fig. 22-3) in layers, leaving leads long enough for a regular connection. The tightness and the fact that they are laid in layers makes the shape proper for twisting. Instead of tying, however, the coils are fastened together by means of tin cleats, taking care not to punch or change the form of the layers of the coil in any way. Four cleats are the usual number needed on a small coil. Once the coil is securely fastened, it may be removed from the loop winder.

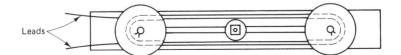

FIG. 22-3 A loop winder is used to obtain uniform coils when form-winding motors.

First, one wooden block is fastened into a bench vise, and the coil is slipped into the groove, with a cover block placed over it. By pulling the second block with both hands, coils are spread to their proper distance. The angles for the coil sides are made by pulling the coil above or below the centerline of the coil while held in the vise. The twist in the ends is made by holding the pliers on the ends of the coils and twisting in the direction of the desired twist, making sure to keep the same twist for both ends.

The twist in the knuckle is usually begun when the coil has been pulled out about halfway. The pliers must be of the flat type so as not to damage the insulation on the wire. The first part of the twist is a half twist, after which the coil is slowly pulled out and the twist is maintained by the pliers. The coil should then be sleeved and taped exactly like the sample coil taken from the motor during stripping.

To secure the coils, bands are normally used. Obtain a pulley, a weight, tin cleats, and bronze or steel wire. The armature stand must be high enough to allow the weight to swing between them. Some means of turning the armature around is also necessary.

Fasten the wire to the armature coil by tape or string and then turn the wire on the armature circumference. The recesses for bands are insulated by a strip of fiber the width of the recess, and then the wire is wound closely together across the area containing the fiber strips. At the same time, place tin cleats under the wire at these points. Wind the wire on loosely so that the wires are close together where they are around the fiber strips and farther apart where they are taken from one strip to the other. Continue this operation until all fiber strips are covered and then, after leaving a little slack, cut the wire off and put the pulley on it and secure. Apply the weight to the pulley, allowing it to hang freely. Next, turn the armature in the direction opposite to that in which it was first turned, at the same time guiding the pulley and wire on the armature so the job will be neat.

When banding an armature, make certain that it lies below the surface of the armature core so the bands will not rub on the stator. Also clean off any excess soldering paste from the band with alcohol so the electrical current will not cause it to corrode.

WINDING STATORS

The diamond coil is also used to wind motor stators and is wound on a loop form as discussed previously. The size and number of wires will depend on the type and size of motor being wound and also on the existing coils in the motor. Therefore, during stripping, one or two sample coils should be saved to use as a guide in winding the new coils.

To better understand the procedures for winding a motor stator, an actual example should be presented.

Let's assume that a stator for a three-phase motor has 36 slots and 6 poles. Applying the formula given earlier in this chapter,

$$\text{coils per group} = \frac{36}{6 \times 3} = 2 \text{ coils/group}$$

The full-pitch coil span can then be found by applying the equation

$$\text{full-pitch coil span} = \sideset{}{}{^{36}\!/_6} + 1 = 7$$

From the preceding, it can be seen that the first coil will span or lie in slots 1 and 7 of the total 36 available.

Once the slots have been insulated as discussed in Chapter 20, begin the winding by placing one side of the first coil in any slot with the leads of the coil toward the winder. One side of the next coil is then placed in the slot to the left of the first, which will make the winding progress in a clockwise direction around the stator. Four more coils are then placed in the slots in a similar manner, leaving the top sides of all of them out.

When the bottom side of the seventh coil is placed in the seventh slot, its top side is laid on top of the first coil, and the bottom of the eighth coil is placed in the eighth slot with its top placed on top of the bottom side of the second coil. See Fig. 22-4. This may be somewhat confusing at first, but if a stator and sample coils are used to demonstrate this procedure, as progress is made, it will quickly become easy to understand.

While the coils of the winding just described were laid in to the left of the first, or clockwise around the stator, they can be laid either clockwise or counterclockwise, according to the shape of the end twist of the coils.

In winding stators of small size, it is usual practice to connect the coils into groups as they are fed in the slots. Note in Fig. 22-4 that the bottom lead of the first coil is connected to the top lead of the second. The top lead of the first coil and the bottom lead of the second coil are identified or marked with sleeving of the same color. All the remaining groups are connected together the same as the first, but the unconnected leads of

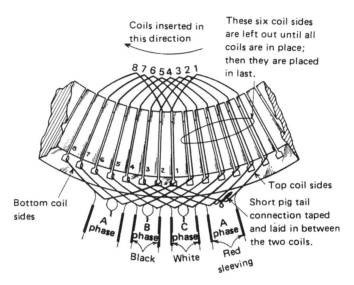

Coils inserted in
this direction

These six coil sides
are left out until all
coils are in place;
then they are placed
in last.

8 7 6 5 4 3 2 1

Top coil sides

Bottom coil
sides

Short pig tail
connection taped
and laid in between
the two coils.

A
phase

B
phase

C
phase

A
phase

Black   White   Red
                           sleeving

FIG. 22-4   Method of placing the first coils in a motor stator and the proper rotation for inserting them. (Courtesy Page Power Co.)

the second group are marked with sleeving color different from the first
and the third group with still another color. For the fourth group, however, the color for the first group is again used, as are the colors for the
other two groups.

After the wedges are in the slots, the pole-group connections are
made as shown in Fig. 22-5. In looking at this diagram, note that the
leads for the A, B, and C phases will show three separate windings spaced
two thirds of a pole, or 120 electrical deg, apart.

A top lead of any convenient coil in the winding is selected for the
start of the A phase, and all groups of a corresponding color are connected into one winding. Then the second start or B phase is selected.
This lead must be taken from the top of the third group, counting the A
phase as number 1. All groups for the B phase are then connected, and
finally those for the C phase are connected. The C phase should start at
the top lead of the fifth coil group, which would be the same distance
from B as B is from A.

There will be six leads left, three starts and three finish leads, all
from the top sides of the coils.

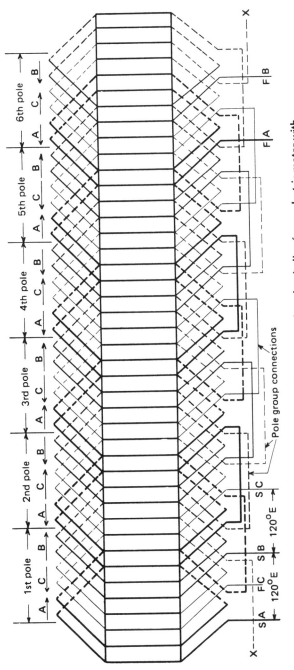

FIG. 22-5 Complete diagram of a three-phase, six-pole winding for an electric motor with 36 slots. (Courtesy Page Power Co.)

23

Rewinding and Reconnecting

At times, electric motors may be rewound to change the operating voltage to permit them to operate on a different line voltage. It may also become necessary to change the phases on a particular motor – usually in an emergency situation.

The voltage of any individual motor winding varies directly with the number of turns it has connected in series. There are, of course, certain practical limits beyond which this change of voltage should not be carried. For example, a motor originally operating at 240 V might possibly be changed – by increasing the number of turns in series – to a point where the winding would stand 2400 V, but it is doubtful whether the insulation would stand so high a voltage.

It is almost always permissible to reconnect a winding to operate on a lower voltage than it has been designed for, but when reconnecting a motor to increase its operating voltage, the insulation should always be considered. The usual ground test for the insulation of such equipment is to apply an alternating-current voltage of twice the machine's rated voltage plus 1000 V. This voltage should be applied from the winding to the frame for at least 1 min, and a test should be made after the winding is reconnected or on any new winding before it is placed in operation. When a winding is reconnected for a different voltage, it should be arranged so that the voltage on each coil group will remain unchanged.

The diagram in Fig. 23-1 shows four coil groups connected in series to accommodate 240 V, which places 60 V on each coil group, causing an

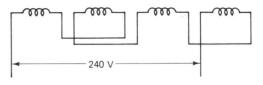

FIG. 23-1 Four coil groups connected
in series to accommodate 240 V.

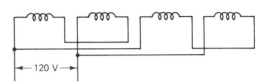

FIG. 23-2 Two coil groups connected
in series in each of the two parallel circuits to accommodate 120 V.

assumed 5 A of current to flow. To be used on a 120-V circuit, the coils
could be arranged as shown in Fig. 23-2. In this arrangement, two coil
groups are in series in each of the two parallel circuits. When 120 V are
applied to these two parallel groups, there will still be 60 V/coil, and the
motor will operate normally on 120 V; even the same amount of current
will flow. The rotating magnetic field will not be affected any differently
as long as the amount of current per coil is not changed and the polarity
of the coils is kept the same.

In connecting three-phase windings, all phases must be connected
for the same number of circuits, and when connecting the groups for a
winding having several circuits, extreme care should be taken to obtain
the correct polarity on each group.

It is common practice for motor manufacturers to design machines
that can readily be operated on either of two common voltages, that is,
120/240, 240/480, etc. This is accomplished by a series or parallel arrangement which can be changed by varying the lead connections to the
motor—usually in the terminal junction box. For example, one three-
phase motor may have one connection where the pole groups in each
phase are connected in series to operate on 480 V. Another connection arrangement for this same motor could have the pole groups for each phase
connected half in series and half in parallel for operation on half the voltage of the first arrangement, or 240 V. In this arrangement, however, to
maintain the same horsepower at half the voltage, the amperes for the
240-V connection would double at full load. This presents no problem
since in the 240-V connection with two circuits in parallel, twice the cross-
sectional area of copper exists than in the series circuit for the 480-V arrangement.

In most dual voltage motors, each winding is divided into two parts
with suitable leads from each section brought outside the motor. These
leads can then be conveniently changed for either one or two voltages.

CHANGE OF PHASES AND FREQUENCY

In certain emergency cases, it is desirable to know how to change a motor from, say, three phase to two phase or vice versa. This can be accomplished by changing the number of coils per groups. As a brief example, a certain two-phase motor having three coils per group could be made to operate on three-phase power by arranging the coils to two per group.

Sometimes it is desirable to change a motor which has been operating on one frequency so that it will operate on a circuit of another frequency. The most common frequency for alternating current in the United States is 60 Hz; in England, 50 Hz; etc. But occasionally (although very infrequently) some other odd frequency may be encountered.

When an induction motor is operating, a rotating magnetic field is set up in the stator, and it is this field which induces the secondary current in the rotor and produces the motor torque. This same rotating field cuts across the coils in the stator itself and generates in them a countervoltage which opposes the applied line voltage and limits the current through the winding. The speed of field rotation governs the strength of the countervoltage (CEMF) and therefore regulates the amount of current which can flow through the winding at any given line voltage.

There are two factors that govern the speed of rotation in this magnetic field. These are the number of poles in the winding and the frequency of the applied alternating current. Any change that is made in the frequency of the current supplied to a motor should be offset by a change of voltage in the same direction and in the same proportion. For example, a motor changed from 30 to 60 Hz should have the magnetic field rotate twice as fast and the CEMF doubled. To maintain the same current value in the stator coils, the line voltage should also be doubled. If the winding is to be operated on the same voltage at this higher frequency, the number of turns in each group across the line should be reduced to one-half the original number to allow the same current to flow.

This procedure should, of course, be reversed when changing a motor to operate on a lower frequency.

When the frequency of a given motor is varied and the stator flux kept constant, the horsepower rating will vary directly with the change in speed. In other words, the horsepower of any motor is proportional to the product of its speed and torque.

CHANGING NUMBER OF POLES AND SPEED

The speed of an induction motor is inversely proportional to the number of poles; that is, if the number of poles is increased to double, the speed will decrease to one half, or if the poles are decreased to one-half their original number, the speed will increase to double.

When changing the number of poles of an induction motor, if the voltage is varied in the same direction and same proportion as the change produced in the speed, the torque will remain practically the same, and the horsepower will vary with the speed. Therefore, the horsepower increases with the higher speeds and decreases at lower speeds in exact proportion to the change of speed.

With induction-type motors, a change in speed invariably involves a change in the number of poles set up by the winding, and since this implies a variation in the coil span, rewinding is usually required.

COMPLETING THE WINDING JOB

Once the windings are positioned in the motor being overhauled, as discussed in previous chapters, the connecting of the leads to the commutator is the next process. When the bottom leads are laid in the slots as the coils are wound, proper sequence is ensured. When the coils are all wound, then the top leads are put in their proper segments—which are adjacent to the bottom leads on lap-wound armatures and with the proper commutator span on wave-wound armatures.

Some armatures are wound completely before the bottom leads are placed into their segments, followed by the top leads. Another method is when the bottom leads are installed through the same slot as the top leads. This type of winding is frequently used on lap-wound armatures where the top and bottom leads swing out of line with the coil itself. Therefore, bringing the bottom leads out of the bottom coil side would necessitate extra long bottom leads. By combining the two leads, the shorter bottom leads will not have a tendency to raise up and cause a short on the revolving armature.

The top and bottom leads can be laid one after the other on some armatures if care is taken to insulate between leads with cotton tape or oiled linen; these leads should also be sleeved. This insulation is interwoven between the leads as shown in Fig. 23-3 and then brought up next to the commutator where the leads need the most insulation because of their closeness.

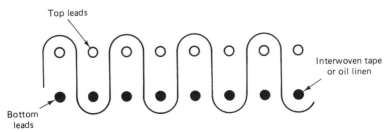

FIG. 23-3 Method of installing interwoven tape between top and bottom leads.

To correctly connect the leads, it is necessary to use a test lamp or continuity meter to perform the following tests. At this point, the commutator bars should have the bottom leads connected to them. Place one test lead onto one commutator bar (with a bottom lead connected to it) and then touch the unconnected top leads with the test lamp or meter until you come to the lead that lights up the lamp or deflects the meter. Connect this lead as shown in the data collected from the motor prior to beginning the overhaul.

For instance, the top lead, if on a lap winding, will be connected to the bar adjacent to that of the bottom lead. If dealing with a wave-wound armature, count the armature span over from the bottom lead to where the top lead connects and put it in the bar. Take the next bar and follow the same procedure and then continue on with each of the succeeding bars until all leads have been connected. Always continue in the direction taken, taking each bar in succession and connecting the top lead for its respective bottom lead.

When checking for top leads of lap-wound armatures with cross connectors in back of the commutator, care must be taken when finding a top lead for a corresponding bottom lead connected in a bar. With the cross connectors there will be two top leads where the lamp lights – since the current travels over the cross connector or jumper – to another bar where there is a bottom lead connected. The current will follow this path and then travel through this coil, having the bottom lead in the same slot as the jumper out to the second top lead. Knowing from the data that this armature is cross-connected, test the leads near the coil from which the bottom lead comes.

When making the final connections to any motor, refer to the data card frequently. Always determine the bottom lead connection for one coil and then lay in the remaining bottom leads in succession as the coils progress.

Where the bottom leads are not placed into the bars as the armature is being wound, find the bottom lead for one coil and place it in its proper slot; then continue as discussed previously. These bottom leads should be marked in any way that is convenient so long as the marking does not injure the insulation in any way.

When the top and bottom leads are installed together, find the bar where the bottom lead of a coil connects and then connect the single lead to a top and bottom lead to that bar. Take the lead from the next coil and connect it to the next bar and so on, being sure to connect the leads to the bars in the same direction as the coil progresses. When two single wires are encountered which are in separate sleeves and are the beginning and ending leads of the winding, connect them to the same bar.

Once the armature has been wound and is ready for connecting, some projects will require some building up between the end of the armature winding and the commutator so the bottom leads will have some means of support. This is accomplished by installing cotton tape around the shaft between the commutator and the winding to fill up the gap. This tape should be wound until it builds up even with the bottom of the slot in the segment. Be sure to keep it as smooth as possible to avoid having to bend the leads out of shape to travel over it. Next, oiled linen or cotton tape should be placed over the winding, covering an area from where the winding leaves the slots to its end so that leads do not rest directly upon it. If leads were allowed to run over the winding with no insulation between, a short circuit might develop due to friction while the motor is running.

The bottom leads are laid in place next, being sure to interweave cotton tape between each lead so they will not cause a short circuit. More cotton tape is used to cover them to prevent the top leads from resting directly upon them. Finally the top leads are placed in their proper position.

Once all leads are in place, all insulation should be scraped off where they make contact with the bars. The leads are then placed in the bars, and any excess lead is cut off. When the bottom leads are placed in the bars first, they are tapped down in the slots with a lead sinker to flatten the lead so it becomes tight in the slot as shown in Fig. 23-4. The top leads are treated in the same manner, being careful in both cases not to crush the leads so they will break off in back of the commutator. Once both leads are in the slot, it is sometimes recommended that the slots be peened over to prevent the leads from rising out. While this peening will secure the leads, it makes them very difficult to remove the next time that the motor needs overhauling.

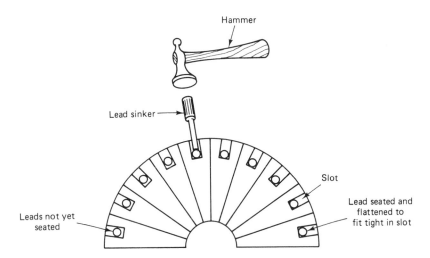

FIG. 23-4   Once leads are installed in the slots, they should be tapped with a lead sinker to slightly flatten them, which will in turn cause them to fit tightly in the slots.

SOLDERING LEADS

Once the leads have been cleaned of insulation and placed in their proper slots, they are ready to be soldered. A temporary protective band should be installed over the leads – especially against the back of the commutator – so when the leads are soldered, none of the solder will flow down in the back of the commutator and short the bars together. A soldering iron or gun is used for the job, but the tips must be well tinned for proper soldering and should not be allowed to become too hot.

Resin-type soldering paste is used to coat the tops of the bars at points where the leads will connect. It is best not to use a paste with an acid base. When the soldering iron is at the right temperature, touch it to the commutator bar and heat until the paste runs down between the leads and the slot in the segment. The bar should be hot enough to allow the solder to take, but be careful not to overheat. Some solder should also be touched against the iron and be allowed to flow into the slot between the leads. Then lift the soldering iron off the bar with a quick jerk as this will leave a smooth finish and will also keep the solder from bridging between commutator bars. Never use so much solder on the iron or the bar to cause the solder to stack up on the top of the bar; not only does this make a bad looking job, but it increases the chances of the solder running over onto the next bar, shorting the two bars together.

Experienced motor repairmen like to allow the soldering gun to be heating part of the next bar as it is heating the bar being worked on at the moment. One way to accomplish this is to use an extralarge soldering iron, one that will produce enough heat to heat the next bar while one is being soldered.

Should the solder run across between the two commutator bars so it covers the mica separating them, it should not be melted off because the bar may become hot enough to cause the solder to run down in back of the connector. It is best to use a hacksaw or a small file to cut the solder over the mica so the mica will again separate the bars so as not to be shorted. Should any solder run down the front of the commutator, this usually presents no problems so long as it does not short two bars together. The commutator may then be turned in a lathe to even the surface.

BAKING AND VARNISHING

All windings, whether dc or ac, should be thoroughly impregnated with a good grade of insulating varnish before they are put into service.

This varnish serves several very important purposes. When properly applied, it penetrates to the inner layers of the coils and acts as extra insulation of the conductors, thereby increasing the dielectric strength of the insulation between them. This compound within the coils and in their outer taping greatly reduces the liability of short circuits between conductors and of grounds to the slots or frame.

When a winding is thoroughly saturated with insulating varnish and this varnish is properly hardened, it adds a great deal to the strength of the coils and holds the conductors rigidly in place. This prevents a great deal of vibration that would otherwise tend to wear and destroy the insulation, particularly in the case of alternating-current windings where the alternating flux tends to vibrate the conductors when in operation.

Insulating varnish also prevents moisture from getting in the coils and reducing the quality of the insulation, and it keeps out considerable dust, dirt, and oil that would otherwise accumulate between the coils. Keeping out moisture, dust, and oil greatly prolongs the life of the insulation.

There are many grades of insulating varnish some of which require baking to *set* or harden them and others which have in them certain liquids or solvents which make them dry and harden very quickly when exposed to air. Varnishes of the first type are called baking varnishes, and the latter are called air-dry varnishes.

Good air-dry insulating varnish will set or harden in from 20 to 30 min, but it should be allowed to dry out thoroughly for about 24 h before the windings are put in service. Air-dry varnish is not considered quite as good as the better grades of baking varnish. Therefore, the latter should be used wherever a bake oven or some means of applying heat is available.

There are several methods used to apply insulating varnish to motor coils and windings. The most popular methods include dipping, brushing, and spraying.

Dipping is considered the best method and should be used for all small windings of stators and armatures and for armatures and stator coils and field coils. To dip these coils or windings, a pan or tank of the proper size and depth will be required. Before dipping the windings, they should be thoroughly dried out in a bake oven at about 212°F in order to drive out all moisture and to heat the coils so that when they are dipped the varnish will rapidly penetrate to their inner layers.

The coils should be allowed to remain in the varnish until all bubbling has ceased. When they seem to have absorbed all the varnish possible, they should be slowly withdrawn from the tank at about the same rate as the varnish flows from them of its own accord. This will give them a uniform coating with the least possible accumulation of varnish at the lower end. They should then be allowed to drain until the varnish stops dripping and becomes partially set. The time required for this will depend on the size of the winding or coils.

When dipping a large number of small coils, considerable time can be saved by arranging a drip board set at an angle, so the coils can be hung above it and the varnish which drips from them will run down the board and back into the tank. With this method other coils can be dipped while the first set is draining.

After all the surplus varnish is drained from the coils, they should be baked. When placing them in the oven, it is a good plan to reverse their positions, so that any excess varnish on the bottom ends will tend to flow back evenly over their surface when first heated.

When applying the varnish with a brush, the winding should, if possible, be preheated to drive out the moisture and permit the varnish to flow deeper into the coils. Varnish can be applied with an ordinary paint brush, and this method is used where the dipping tank is not large enough to accommodate the winding or where no dipping tank is available.

Spraying is used principally on large windings and gives a very good surface for a finishing coat.

The ends of coils should be given two or three coats of varnish as an added protection against mechanical damage and moisture and to help prevent flashovers to the frame of the machine.

Try not to get insulating varnish on the commutator, as it makes it hard to turn. Allow the armature to drain for a few minutes and then clean off all excess varnish from the commutator, shafts, and armature insulations with gasoline or AWA 1,1,1. It is necessary that all varnish be taken off the laminations so the armature will turn freely between the fields or stator. It must be taken off the shaft so oil may properly lubricate. When this is done, put the armature back into the oven and bake it for a few hours, being careful not to get it too hot.

Manufacturers of insulating varnish usually furnish convenient tables with their product which give the proper temperatures and approximate time for baking. Be aware that when baking complete armature windings, more time is required to thoroughly bake the larger sizes. As a rule, slower baking produces a more elastic and better-quality insulation than when baking quickly at higher temperatures.

Besides the insulating qualities, insulating varnish also provides a smoother surface on the windings and coils, making them much easier to clean either by means of a brush, compressed air, or washing them with some degreasing solutions such as AWA 1,1,1.

24

Motor Efficiency

Electric motors, which consume a major portion of the electrical energy generated in the United States, are nothing more than converters. They convert electrical energy to mechanical energy, which is used by the machine to perform work. Modern well designed industrial motors convert about 85% to 95% of the electrical power to usable mechanical energy. The real potential to conserve energy exists with the equipment being driven and not the motor that is doing the work.

The efficiency of an electric motor is the ratio of the electrical energy in to the mechanical energy out. It is expressed by the equation:

$$Efficiency\ (\%) = \frac{0.746 \times Mechanical\ Power\ Out}{Electrical\ Power\ In} \times 100$$

Where 0.746 is a constant that converts horsepower to kilowatts. Multiplication by 100 represents the results in percent.

Note that the metric system designation for motors defines size by kilowatts rather than horsepower. For example, a 10 hp motor is a 7.5 (7.46) kW motor in the metric system.

The efficiency of standard 5 hp and larger motors is in the range of 85% to 95%. Motors in this size range account for most of the total motor energy consumed. The lower efficiencies of the smaller motors (which may be as low as 75% for a 1 hp motor) are of no great concern in any real effort to conserve energy since their total energy consumption is so small.

Efficiencies vary only slightly (typically 1 to 1.5 points) over the motor's normal operating range of half load to full load. Furthermore, efficiency typically peaks at ¾ load.

The electrical energy consumed — the 5% to 15% of the total energy input depending on motor size — by the motor to perform its function might be considered the cost to convert electrical energy into mechanical energy and is defined by the motor design engineer as the watts loss of the motor. It consists of such factors as FR losses in the stator and rotor, bearing friction, windage, etc. Watts loss is a function of motor design and construction. Such factors as lamination steel, stack length, air gap, rotor casting methods, ventilating air flow and bearing design affect watts loss. Consequently, these factors are involved in any engineering effort to improve a motor's efficiency. Improvement beyond an optimum level, however, tends to cause excessive increase in manufacturing costs and motor costs.

Conservative design objective for standard motors is to deliver a balance that maximizes efficiency while minimizing manufacturing costs.

DETERMINING MOTOR EFFICIENCY

In recent years, *NEMA* established standard MG 1-12.53a defining motor efficiency factors. It requires that testing be done in accordance with *IEEE* Standard 112. For accuracy, consistency and comparability, the values reported by various motor manufacturers should reference the *NEMA* Standard and *IEEE*-112 Test Method B. The testing techniques require using dynamometers sized to the motor and carefully calibrated instrumentation. As a result, customer validation of a motor's efficiency is costly and in most cases impractical.

Because of variances existing in manufacturing, motors built to the same manufacturing specification in the same production run will vary from motor to motor. Therefore, the efficiency figure on the nameplate is a nominal (average) value. As indicated in the following example, electively reproduced from Table 12-4 in *NEMA* Standard MG 1-12.53b, the Standard requires a guaranteed minimal efficiency value for a nominal nameplate as follows:

Nominal Efficiency	Guaranteed Minimal Efficiency
95.0	94.1
93.0	91.7
90.2	88.5
85.5	82.5

Notice that this Standard does not relate horsepower to these efficiency values and, in fact, codes or specifications requiring specific efficiencies for certain size motors do not exist. Some companies and some engineering firms, however, have established specifications for equipment they design, sell or use.

Users must rely on motor manufacturer ratings for the efficiency figures they need to calculate energy and dollar savings. Since nameplate efficiencies are nominal ratings and not the actual value of the motor, however, an exact savings analysis can not be made.

CALCULATING SAVINGS

Saving electrical energy reduces the power bill. However, a sensible energy cost savings evaluation must equate the energy savings with the price premium usually associated with the "energy efficient" motor. Annual dollar savings can be easily calculated using the equation:

$Savings/yr =

$$0.746 \times hp \times \left(\frac{100}{Eff_A} - \frac{100}{Eff_B} \right) \times hr \times \$/kWh$$

Where:

hp	=	(typical or average) load on the motor
Eff_A	=	the existing motor;
Eff_B	=	the replacement motor
Hours (hr)	=	annual hours of operation under the specific horsepower load*
*/kWh	=	average cost of electricity per kilowatt hour*

*4000 hours per year, average power cost $.06/kWh.

Operating load and time affect savings substantially. A motor loaded to half its rated load uses only half the full load input power and the savings can be only half as great. A motor operated on a two shift, five day schedule (4000 hours/year) uses twice as much power as the same motor operating on one shift. Meaningful calculations must reflect an effort to incorporate typical average values for these variables.

Once the annual savings are known the time needed for the savings to cover the motor price premium can be calculated using the equation:

$$Payback\ Period = \frac{Price\ Premium}{Yearly\ Savings}$$

POWER FACTOR AND ENERGY CONSUMPTION

The impedance of an electric motor (and other AC equipment) causes the current to lag behind the voltage at some angle θ. This lagging current, I_m is the meter read current, and it is the current drawn by the motor. Energy consumed, however, is determined by the current in phase with the voltage, I_e. As shown in following sample vector diagrams (see Fig. 24-1), I_m decreases, but I_c stays the same with improving power factor (as angle θ decreases). By definition, power factor (pf) is the cosine of the angle θ. Therefore,

$$I_e = I_m \times \cos \theta = I_m \times pf.$$

In a three-phase system, the equation to compute the energy or kilowatts used by the motor is:

$$Kw = \frac{\sqrt{3} \times E \times I_m \times pf}{1000}$$

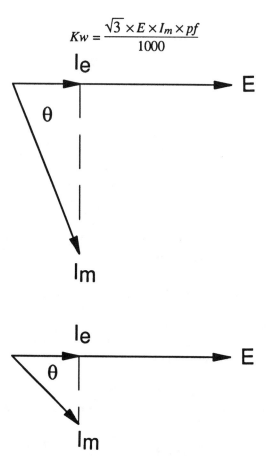

FIG. 24-1 By definition, power factor (pf) is the cosine of angle θ. Therefore, $I_e = I_m \times \cos \theta = I_m \times pf$.

Where:

E = Voltage.

To illustrate, the load current of a typical 5 hp motor operating at 240 V with or without capacitors, is reduced as follows:

	pf	**Load Current, Amps**	**pf Correction**
Old	.82	13.6	None
New	.95	11.7	Capacitors

The power consumption for these two motor is:

$$No\ pf\ Corr. = \frac{\sqrt{3} \times 240 \times 13.6 \times .82}{1000} = 4.4\ kW$$

$$With\ pf\ Corr. = \frac{\sqrt{3} \times 240 \times 11.7 \times .95}{1000} = 4.4\ kW$$

Therefore, improved power factor does *not* save energy.

POWER FACTOR AND ENERGY COSTS

Some power companies apply a power factor or "reactive demand" charge to the power bill of industrial customers. The impact of this charge depends directly upon the customer's power factor. It typically ranges from 0.5% to 2.5% of the total power bill and is inconsequential when seeking important cost savings to offset the premium price of power factor corrected motors or power factor correction devices.

To reduce the reactive demand charge with capacitors, a single installation to correct the entire system rather than separate correction capacitors at individual motors is the most economical, and usually adequate solution.

POWER FACTOR AND BUILDING CODES

Section 8.2 of the now obsolete ASHRAE Standard 90.75 stated that motors used in new building construction shall have a power factor of 0.85 or be corrected to 0.90 or better. Unfortunately, this Standard has been incorporated into the building codes of almost all states as an energy conservation measure. The *ANSI/AS-HRAE/IES* Standard for Energy Conservation in New Building Design — 90A-

1980 — has been adopted as a revision to the old *ASHRAE* Standard 90.75. This revision recognizes that using power factor to attempt to conserve energy is an error, and the requirement for a power factor of 0.85, or corrected to 0.90 or better, has been eliminated from the revised standard. Unfortunately the state building codes have yet to be modified to reflect this revision. Nevertheless, it is logical that implementation of this section of the state codes can be delayed pending similar action by the state legislatures.

NEMA has recently adopted into the MG-1 Standard a table of Nominal and Minimum Efficiency values for energy efficient motors. These tables are published in full in Figs. 24-2 and 24-3.

Note: A deep and grateful bow is made in the direction of Jean Revelt and The Lincoln Electric Co., Cleveland, Ohio for supplying reference material used in this chapter.

HP	2 POLE Nominal Efficiency	2 POLE Minimum Efficiency	4 POLE Nominal Efficiency	4 POLE Minimum Efficiency	6 POLE Nominal Efficiency	6 POLE Minimum Efficiency	8 POLE Nominal Efficiency	8 POLE Minimum Efficiency
1.0	—	—	82.5	80.0	77.0	74.0	72.0	68.0
1.5	80.0	77.0	82.5	80.0	82.5	80.0	75.5	72.0
2.0	82.5	80.0	82.5	80.0	84.0	81.5	85.5	82.5
3.0	82.5	80.0	86.5	84.0	85.5	82.5	86.5	84.0
5.0	85.5	82.5	86.5	84.0	86.5	84.0	87.5	85.5
7.5	85.5	82.5	88.5	86.5	88.5	86.5	88.5	86.5
10.0	87.5	85.5	88.5	86.5	90.2	88.5	89.5	87.5
15.0	89.5	87.5	90.2	88.5	89.5	87.5	89.5	87.5
20.0	90.2	88.5	91.0	89.5	90.2	88.5	90.2	88.5
25.0	91.0	89.5	91.7	90.2	91.0	89.5	90.2	88.5
30.0	91.0	89.5	91.7	90.2	91.7	90.2	91.0	89.5
40.0	91.7	90.2	92.4	91.0	91.7	90.2	90.2	88.5
50.0	91.7	90.2	92.4	91.0	91.7	90.2	91.7	90.2
60.0	93.0	91.7	93.0	91.7	92.4	91.0	92.4	91.0
75.0	93.0	91.7	93.6	92.4	93.0	91.7	93.6	92.4
100.0	93.0	91.7	93.6	92.4	93.6	92.4	93.6	92.4
125.0	93.0	91.7	93.6	92.4	93.6	92.4	93.6	92.4
150.0	93.6	92.4	94.1	93.0	93.6	92.4	93.6	92.4
200.0	93.6	92.4	94.1	93.0	94.1	93.0	93.6	92.4

FIG. 24-2 Full load efficiencies for open motors.

HP	2 POLE Nominal Efficiency	2 POLE Minimum Efficiency	4 POLE Nominal Efficiency	4 POLE Minimum Efficiency	6 POLE Nominal Efficiency	6 POLE Minimum Efficiency	8 POLE Nominal Efficiency	8 POLE Minimum Efficiency
1.0	—	—	80.0	77.0	75.5	72.0	72.0	68.0
1.5	78.5	72.5	81.5	78.5	82.5	80.0	75.5	72.0
2.0	81.5	78.5	82.5	80.0	82.5	80.0	82.5	80.0
3.0	82.5	80.0	84.0	81.5	84.0	81.5	81.5	78.5
5.0	85.5	82.5	85.5	82.5	85.5	82.5	84.0	81.5
7.5	85.5	82.5	87.5	85.5	87.5	85.5	85.5	82.5
10.0	87.5	85.5	87.5	85.5	87.5	85.5	87.5	85.5
15.0	87.5	85.5	88.5	86.5	89.5	87.5	88.5	86.5
20.0	88.5	86.5	90.2	88.5	89.5	87.5	89.5	87.5
25.0	89.5	87.5	91.0	89.5	90.2	88.5	89.5	87.5
30.0	89.5	87.5	91.0	89.5	91.0	89.5	90.2	88.5
40.0	90.2	88.5	91.7	90.2	91.7	90.2	90.2	88.5
50.0	90.2	88.5	92.4	91.0	91.7	90.2	91.0	89.5
60.0	91.7	90.2	93.0	91.7	91.7	90.2	91.7	90.2
75.0	92.4	91.0	93.0	91.7	93.0	91.7	93.0	91.7
100.0	93.0	91.7	93.6	92.4	93.0	91.7	93.0	91.7
125.0	93.0	91.7	93.6	92.4	93.0	91.7	93.6	92.4
150.0	93.0	91.7	94.1	93.0	94.1	93.0	93.6	92.4
200.0	94.1	93.0	94.5	93.6	94.1	93.0	94.1	93.0

FIG. 24-3 Full load efficiencies for enclosed motors.

Appendix: Trade Sources

MOTORS AND MOTOR CONTROLS

• Contractors

Allen Bradley Co.
1201 S. Second St.
Milwaukee, WI 53204

Arrow-Hart Inc.
Division of Crouse-Hinds
103 Hawthorn St.
Hartford, CT 06101

Automatic Switch Co.
50-56 Hanover Rd.
Florham Park, NJ 07932

Crydom
1521 Grand Av.
El Segundo, CA 90245

Cutler-Hammer Inc.
4201 N. 27th St.
Milwaukee, WI 53216

Duraline
Division of J.B. Nottingham & Co.,
  Inc.
75 Hoffman Ln.
Central Islip, NY 11722

Essex Group
131 Godfrey
Logansport, IN 46947

Federal Pacific Electric Co.
150 Av. L
Newark, NJ 07101

Furnas Electric Co.
1000 McKee St.
Batavia, IL 60510

General Electric Co.
General Purpose Control Dept.
P.O. Box 913
Bloomington, IL 61701

General Electric Co.
Industrial Control Dept.
1501 Roanoke Blvd.
Salem, VA 24153

Gould, Inc.
I-T-E Electrical Products
Rollings Meadows, IL 60008

GTE Sylvania, Inc.
Electrical Equipment Group
One Stamford Forum
Stamford, CT 06904

H-B Instrument Co.
4314 N. American St.
Philadelphia, PA 19140

Klockner-Moeller Corp.
Motor Controls
4 Strathmore Rd.
Natick, MA 01760

Mack Electric Devices Inc.
211 Glenside Av.
Wyncote, PA 19095

Magnecraft Electric Co.
5575 N. Lynch
Chicago, IL 60630

Payne Engineering Co.
Box 70
Scott Depot, WV 25560

Ross Engineering Corp.
559 Westchester Dr.
Campbell, CA 95008

Siemens-Allis
Industrial Controls Division
P.O. Box 89
Wichita Falls, TX 76307

Square D Company
P.O. Box 472
Milwaukee, WI 53201

Struthers-Dunn, Inc.
Lambs Rd.
Pitman, NJ 08071

Sylvania Electrical Control
Electrical Equipment Group
One Stamford Forum
Stamford, CT 06904

Telemecanique Inc.
2625 S. Clearbrook Dr.
Arlington Heights, IL 60005

Vectrol Inc.
110 Douglas Rd.
Oldsmar, FL 33557

Ward Leonard Electric Co., Inc.
The Unimax Group
31 South St.
Mt. Vernon, NY 10550

Westinghouse
Control Products Division
Tuscarawas Rd.
Beaver, PA 15009

Zenith Controls, Inc.
830 W. 40th St.
Chicago, IL 60609

TESTING AND MEASURING DEVICES (INSTRUMENTS)

• Ammeters
AEMC Corp.
No. Amer. Dist. Chauvin Arnoux
  Prod.
729 Boylston St.
Boston, MA 02116

Amprobe Instrument
Division of Core Industries Inc.
630 Merrick Rd.
Lynbrook NY 11563

(F.W.)Bell, Inc.
Arnold Engineering Co.
4949 Freeway Dr. E
Columbus, OH 43229

Columbia Electric Mfg. Co.
4519 Hamilton Av.
Cleveland, OH 44114

Control Power Systems Inc.
18978 NE 4th Ct.
North Miami Beach, FL 33179

Etcon Corp.
12243 S. 71st Av.
Palos Heights, IL 60463

General Electric Co.
Instrument Products Operation
40 Federal St.
Lynn, MA 01910

General Electric Co.
Instrument Rental Program
1 River Rd.
Schenectady, NY 12345

Hickok Electrical Instruments
10541 Dupont Av.
Cleveland, OH 44108

Hioki New York Corp.
46-16 235th St.
Douglaston, NY 11363

Martindale Electric Co.
1375 Hird Av.
Cleveland, OH 44107

Pacer Industries Inc.
704 E. Grand Av.
Chippewa Falls, WI 54729

RFL Industries Inc.
Powerville Rd.
Boonton, NJ 07005

Sangamo Weston, Inc.
Schlumberger Division
P.O. Box 3347
Springfield, IL 62714

Simpson Electric Co.
Division of American Gage and
   Machine Co.
853 Dundee Av.
Elgin, IL 60120

Snap-On Tools Corp.
2801 80th St.
Kenosha, WI 53140

(A.W.) Sperry Instruments Inc.
245 Marcus Bl.
Hauppauge, NY 11787

Square D Company
P.O. Box 6440
Clearwater, FL 33518

(H.H.) Sticht Co., Inc.
27 Park Pl.
New York, NY 10007

TIF Instruments Inc.
3661 NW 74th St.
Miami, FL 33147

Triplett Corp.
Bluffton, OH 45817

Western Electro Mechanical
300 Broadway
Oakland, CA 94607

Westinghouse
Relay-Instrument Division
95 Orange St., P.O. Box 606
Newark, NJ 07101

Weston Instruments
A Division of Sangamo Weston Inc.
614 Frelinghuysen Av.
Newark, NJ 07114

Yokogawa Corp. of America
5 Westchester Plaza
Elmsford, NY 10523

• **Cable Tracers**

Amprobe Instrument
Division of Core Industries Inc.
630 Merrick Rd.
Lynbrook, NY 11563

Aqua-Tronics Inc.
17040 SW Shaw St.
Beaverton, OR 97005

Associated Research Inc.
8221 N. Kimball Av.
Skokie, IL 60076

(James G.) Biddle Co.
Township Line & Jolly Rds.
Plymouth Meeting, PA 19462

Cranleigh Electro-Thermal Inc.
P.O. Box 7500
Menlo Park, CA 94025

FRL Inc.
Fisher Div.
517 Marine View Av.
Belmont, CA 94002

General Electric Co.
Instrument Rental Program
1 River Rd.
Schenectady, NY 12345

Goldak Co., Inc.
626 Sonora Av.
Glendale, CA 91201

Hipotronics
Route 22
Brewster, NY 10509

Metrotech Corp.
670 National Av.
Mountain View, CA 94043

Radar Engineers
Division of Epic Corp.
4654 NE Columbia Blvd.
Portland, OR 97218

Rycom Instruments Inc.
9351 E. 59th St.
Raytown, MO 64133

Systems Research, Inc.
P.O. Box 25280
Portland, OR 97225

TIF Instruments Inc.
3661 NW 74th St.
Miami, FL 33147

Tinker & Rasor
P.O. Box 281
San Gabriel, CA 91778

Utility Products
Division of Maxwell Laboratories
    Inc.
8835 Balboa Av.
San Diego, CA 92123

Western Progress
835 Maude Av.
Mountain View, CA 94043

• **Circuit Breaker Testers**
Anderson Power Products Inc.
145 Newton St.
Boston, MA 02135

(James G.) Biddle Co.
Township Line & Jolly Rds.
Plymouth Meeting, PA 19462

Electroware Products Inc.
24 Lisa Dr.
Dix Hills, NY 11746

General Electric Co.
Instrument Rental Program
1 River Rd.
Schenectady, NY 12345

Hioki New York Corp.
46-16 235th St.
Douglaston, NY 11363

Hipotronics, Inc.
Route 22
Brewster, NY 10509

Leviton Mfg. Co., Inc.
59-25 Little Neck Pwy.
Little Neck, NY 11362

TIF Instruments Inc.
3661 NW 74th St.
Miami, FL 33147

• **Continuity Testers**
AEMC Corp.
No. Amer. Dist. Chauvin Arnoux
    Prod.
729 Boylston St.
Boston, MA 02116

Amprobe Instrument
Division of Core Industries Inc.
630 Merrick Rd.
Lynbrook, NY 11563

Associated Research Inc.
8221 N. Kimball Av.
Skokie, IL 60076

Beckman Research & Mfg. Corp.
111 W. Ash Av.
Burbank, CA 91502

(James G.) Biddle Co.
Township Line & Jolly Rds.
Plymouth Meeting, PA 19462

(W.H.) Brady Co.
Division of Industrial Products
2221 W. Camden Rd.
Milwaukee, WI 53201

Bright Star Industries Inc.
600 Getty Av.
Clifton, NJ 07015

Burnworth Tester Co.
815 Ponoma Av.
El Cerrito, CA 94530

Control Power Systems Inc.
18978 NE 4th Ct.
North Miami Beach, FL 33179

Ecos Electronics Corp.
205 W. Harrison St.
Oak Park, IL 60304

Electroware Products Inc.
24 Lisa Dr.
Dix Hills, NY 11746

Etcon Corp.
12243 S. 71st Av.
Palos Heights, IL 60463

Ft. Wayne Electrical Center
Division of Weingart Inc.
1800 Broadway
Ft. Wayne, IN 46804

General Electric Co.
Instrument Products Operation
40 Federal St.
Lynn, MA 01910

Genisco Tech. Corp.
18435 Susana Rd.
Compton, CA 90221

Hipotronics Inc.
Route 22
Brewster, NY 10509

Ideal Industries, Inc.
5224 Becker Pl.
Sycamore, IL 60178

ITT Holub Industries
443 Elm St.
Sycamore, IL 60178

Martindale Electric Co.
1375 Hird Av.
Cleveland, OH 44107

Paragon Electric Co. Inc.
AMF Incorporated
606 Parkway Blvd.
Box 28
Two Rivers, WI 54241

Peschel Instruments Inc.
1412 Viscaya Pkwy.
Cape Coral, FL 33904

Ric-Nor Co. Inc.
10 Roland St.
Charlestown, MA 02129

Simpson Electric Co.
Division of American Gage and
   Machine Co.
853 Dundee Av.
Elgin, IL 60120

Snap-On Tools Corp.
2801 80th St.
Kenosha, WI 53140

(A.W.) Sperry Instruments, Inc.
245 Marcus Bl.
Hauppauge, NY 11787

(H.H.) Sticht Co., Inc.
27 Park Pl.
New York, NY 10007

Systems Research, Inc.
P.O. Box 25280
Portland, OR 97225

Teledyne Penn-Union Monarch Fuse
Division of Teledyne Inc.
229 Waterford St.
Edinboro, PA 16412

Techni-Tool Inc.
Apollo Rd.
Plymouth Meeting, PA 19462

TIF Instruments, Inc.
3661 NW 74th St.
Miami, FL 33147

Triplett Corp.
Bluffton, OH 45817

Vaco Products Co.
1510 Skokie Blvd.
Northbrook, IL 60062

Woodhead Co., Daniel
Division of Daniel Woodhead Inc.
3411 Woodhead Dr.
Northbrook, IL 60062

• Dynamometers

(W.C.) Dillon & Co., Inc.
14620 Keswick St.
Van Nuys, CA 91407

General Electric Co.
Direct Current Motor & Generator
   Dept.
3001 E. Lake Rd.
Erie, PA 16531

Linemen's Supply
Division of Buckingham Mfg. Co.,
   Inc.
7-9 Travis Av.
Binghamton, NY 13904

Simpson Electric Co.
Division of American Gage and
   Machine Co.
853 Dundee Av.
Elgin, IL 60120

Techni-Tool Inc.
Apollo Rd.
Plymouth Meeting, PA 19462

Weston Instruments
A Division of Sangamo Weston Inc.
614 Frelinghuysen Av.
Newark, NJ 07114

• Fault Indicators, URD

Associated Research Inc.
8221 N. Kimball Av.
Skokie, IL 60076

Burndy Corp.
Richards Av.
Norwalk, CT 06856

(James G.) Biddle Co.
Township Line & Jolly Rds.
Plymouth Meeting, PA 19462

Ecos Electronics Corp.
205 W. Harrison St.
Oak Park, IL 60304

Genisco Technology Corp.
18435 Susana Rd.
Compton, CA 90221

Kabo Electronics
123 Bacon
Natick, MA 01760

• Fault Locators
AEMC Corp.
No. Amer. Dist. Chauvin Arnoux
 Prod.
729 Boylston St.
Boston, MA 02116

Aqua-Tronics Inc.
17040 SW Shaw St.
Beaverton, OR 97005

Associated Research Inc.
8221 N. Kimball Av.
Skokie, IL 60076

(James G.) Biddle Co.
Township Line & Jolly Rds.
Plymouth Meeting, PA 19462

Ecos Electronics Corp.
205 W. Harrison St.
Oak Park, IL 60304

Fisher Research Laboratory
Fisher Division
517 Marine View Av.
Belmont, CA 94002

Goldak Co., Inc.
626 Sonora Av.
Glendale, CA 91201

Hipotronics Inc.
Route 22
Brewster, NY 10509

Kabo Electronics
123 Bacon
Natick, MA 01760

Peschel Instruments Inc.
1412 Viscaya Pkwy.
Cape Coral, FL 33904

Progressive Electronics Inc.
432 S. Extension Rd.
Mesa, AZ 85202

Rycom Instruments Inc.
9351 E. 59th St.
Raytown, MO 64133

Sotcher Measurement Inc.
1120 J. Stewart Court
Sunnyvale, CA 94086

Square D. Company
P.O. Box 6440
Clearwater, FL 33518

Systems Research, Inc.
P.O. Box 25280
Portland, OR 97225

TIF Instruments, Inc.
3661 NW 74th St.
Miami, FL 33147

Utility Products
Division of Maxwell Laboratories,
 Inc.
8835 Balboa Av.
San Diego, CA 92123

(The) Von Corp.
P.O. Box 3566G
Birmingham, AL 35211

Western Progress
835 Maude Av.
Mountain View, CA 94043

• GFI Circuit Testers
Daltec Systems Inc.
Box 157 (Onondaga Br.)
Syracuse, NY 13215

Ecos Electronics Corp.
205 W. Harrison St.
Oak Park, IL 60304

Etcon Corporation
12243 S. 71st Av.
Palos Heights, IL 60463

Genisco Technology Corp.
18435 Susana Rd.
Compton, CA 90221

Harvey Hubbell Inc.
Wiring Device Division
State St.
Bridgeport, CT 06602

Kabo Electronics
123 Bascon
Natick, MA 01760

Leviton Mfg. Co. Inc.
59-25 Little Neck Pwy
Little Neck, NY 11362

Sotcher Measurement Inc.
1120 J Stewart Court
Sunnyvale, CA 94086

Triple S Products
200 E. Prairie St.
Vicksburg, MI 49097

Wadsworth Electric Mfg. Co. Inc.
P.O. Box 272
Covington, KY 41012

• Ground Detectors

Associated Research Inc.
8221 N. Kimball Av.
Skokie, IL 60076

Burnworth Tester Co.
815 Pomona Av.
El Cerrito, CA 94530

Eagle Electric Mfg. Co. Inc.
45-31 Court Square
Long Island City, NY 11101

Ecos Electronics Corp.
205 W. Harrison St.
Oak Park, IL 60304

Erickson Electrical Equipment Co.
4460 N. Elston Av.
Chicago, IL 60630

General Electric Co.
Instrument Rental Program
1 River Rd.
Schenectady, NY 12345

Hipotronics Inc.
Route 22
Brewster, NY 10509

Kabo Electronics
123 Bacon
Natick, MA 01760

Key Systems Inc.
Allenwood-Herbertsville Rd.
Howell, NJ 07731

Midland-Ross Corp.
Electrical Products Div.
530 W. Mt. Pleasant Av.
Livingston, NJ 07039

Rochester Instrument Systems Inc.
255 N. Union St.
Rochester, NY 14605

Russellstoll
Electrical Products Division
Midland-Ross Corp.
530 W. Mt. Pleasant Av.
Livingston, NJ 07039

Sotcher Measurement Inc.
1120 J. Stewart Court
Sunnyvale, CA 94086

Wadsworth Electric Mfg. Co. Inc.
P.O. Box 272
Covington, KY 41012

Westinghouse
Relay-Instrument Division
95 Orange St.
P.O. Box 606
Newark, NJ 07101

• Ground Testers

AEMC Corporation
No. Amer. Dist. Chauvin Arnoux
 Prod.
729 Boylston St.
Boston, MA 02116

AMP Special Industries
Division of AMP Products Corp.
Valley Forge, PA 19482

Associated Research Inc.
8221 N. Kimball Av.
Skokie, IL 60076

Burnworth Tester Co.
815 Pomona Av.
El Cerrito, CA 94530

Eagle Electric Mfg. Co., Inc.
45-31 Court Square
Long Island City, NY 11101

Ecos Electronics Corp.
205 W. Harrison St.
Oak Park, IL 60304

General Electric Co.
Instrument Rental Program
1 River Rd.
Schenectady, NY 12345

Genisco Technology Corp.
18435 Susana Rd.
Compton, CA 90221

Hioki New York Corp.
46-16 235th St.
Douglaston, NY 11363

Hipotronics, Inc.
Route 22
Brewster, NY 10509

Independent Protection Co. Inc.
1603-09 S. Main St.
Goshen, IN 46526

Ric-Nor Co., Inc.
10 Roland St.
Charlestown, MA 02129

Russellstoll
Division of Electrical Products
Midland-Ross Corp.
530 W. Mt. Pleasant Av.
Livingston, NJ 07039

Simpson Electric Co.
American Gage and Machine Co.
  Division
853 Dundee Av.
Elgin, IL 60120

Sotcher Measurement Inc.
1120 J. Stewart Court
Sunnyvale, CA 94086

(H.H.) Sticht Co., Inc.
27 Park Pl.
New York, NY 10007

TIF Instruments Inc.
3661 NW 74th St.
Miami, FL 33147

Triplett Corp.
Bluffton, OH 45817

(Daniel) Woodhead Co.
Division of Daniel Woodhead Inc.
3411 Woodhead Dr.
Northbrook, IL 60062

Yokogawa Corp. of America
5 Westchester Plaza
Elmsford, NY 10523

• **Insulation Testers**
AEMC Corp.
No. Amer. Dist. Chauvin Arnoux
  Prod.
729 Boylston St.
Boston, MA 02116

Amprobe Instrument
Division of Core Industries Inc.
630 Merrick Rd.
Lynbrook, NY 11563

Associated Research Inc.
8221 N. Kimball Av.
Skokie, IL 60076

Beckman Instruments Inc.
Division of Cedar Grove Operations
89 Commerce Rd.
Cedar Grove, NJ 07009

(James G.) Biddle Co.
Township Line & Jolly Rds.
Plymouth Meeting, PA 19462

Burnworth Tester Co.
815 Pomona Av.
El Cerrito, CA 94530

Ecos Electronics Corp.
205 W. Harrison St.
Oak Park, IL 60304

Ft. Wayne Electrical Center
Division of Weingart, Inc.
1800 Broadway
Ft. Wayne, IN 46804

Hioki New York Corp.
46-16 235th St.
Douglaston, NY 11363

Hipotronics, Inc.
Route 22
Brewster, NY 10509

Martindale Electric Co.
1375 Hird Av.
Cleveland, OH 44107

Peschel Instruments Inc.
1412 Viscaya Pkwy.
Cape Coral, FL 33904

Ricca-Reddington Instruments, Inc.
1400G NW 65th Av.
Plantation, FL 33313

Ross Engineering Corp.
559 Westchester Dr.
Campbell, CA 95008

Simpson Electric Co.
Division of American Gage and
  Machine Co.
853 Dundee Av.
Elgin, IL 60120

Sotcher Measurement Inc.
1120 J Stewart Court
Sunnyvale, CA 94086

(A.W.) Sperry Instruments Inc.
245 Marcus Bl.
Hauppauge, NY 11787

(H.H.) Sticht Co., Inc.
27 Park Pl.
New York, NY 10007

TIF Instruments, Inc.
3661 NW 74th St.
Miami, FL 33147

(The) Von Corp.
P.O. Box 3566G
Birmingham, AL 35211

(Daniel) Woodhead Co.
Division of Daniel Woodhead Inc.
3411 Woodhead Dr.
Northbrook, IL 60062

Yokogawa Corp. of America
5 Westchester Plaza
Elmsford, NY 10523

• **Line Testers**
Beckman Research & Mfg. Corp.
111 W. Ash Av.
Burbank, CA 91502

(James G.) Biddle Co.
Township Line & Jolly Rds.
Plymouth Meeting, PA 19462

(W.H.) Brady Co.
Industrial Products Division
2221 W. Camden Rd.
Milwaukee, WI 53201

Burnworth Tester Co.
815 Pomona Av.
El Cerrito, CA 94530

Ecos Electronics Corp.
205 W. Harrison St.
Oak Park, IL 60304

Etcon Corporation
12243 S. 71st Av.
Palos Heights, IL 60463

Electroware Products Inc.
24 Lisa Dr.
Dix Hills, NY 11746

Gardner Bender, Inc.
6101 N. Baker Rd.
P.O. Box 23322
Milwaukee, WI 53209

General Electric Co.
Instrument Products Operation
40 Federal St.
Lynn, MA 01910

Genisco Technology Corp.
18435 Susana Rd.
Compton, CA 90221

Hipotronics, Inc.
Route 22
Brewster, NY 10509

Pacer Industries Inc.
704 E. Grand Av.
Chippewa Falls, WI 54729

Progressive Electronics Inc.
432 S. Extension Rd.
Mesa, AZ 85202

Ric-Nor Co., Inc.
10 Roland St.
Charlestown, MA 02129

Rodale Mfg./Square D Co.
Sixth & Minor Sts.
Emmaus, PA 18049

TIF Instruments Inc.
3661 NW 74th St.
Miami, FL 33147

Time Mark Corp.
P.O. Box 15127
Tulsa, OK 74115

Triplett Corp.
Bluffton, OH 45817

Vaco Products Co.
1510 Skokie Blvd.
Northbrook, IL 60062

• **Measuring Wheels**
Bessemer Manufacturing Corp.
412 W. King St.
York, PA 17405

(James G.) Biddle Co.
Township Line & Jolly Rds.
Plymouth Meeting, PA 19462

Linemen's Supply
Division of Buckingham Mfg. Co.,
   Inc.
7-9 Travis Av.
Binghamton, NY 13904

Rolatape Corp.
4221 Redwood Av.
Los Angeles, CA 90066

• **Megohmmeters**
AEMC Corp.
No. Amer. Dist. Chauvin Arnoux
   Prod.
729 Boylston St.
Boston, MA 02116

Amprobe Instrument
Division of Core Industries Inc.
630 Merrick Rd.
Lynbrook, NY 11563

Associated Research Inc.
8221 N. Kimball Av.
Skokie, IL 60076

Beckman Instruments Inc.
Division of Cedar Grove Operations
89 Commerce Rd.
Cedar Grove, NJ 07009

Beckman Research & Mfg. Corp.
111 W. Ash Av.
Burbank, CA 91502

(James G.) Biddle Co.
Township Line & Jolly Rds.
Plymouth Meeting, PA 19462

Ecos Electronics Corp.
205 W. Harrison St.
Oak Park, IL 60304

General Electric Co.
Instrument Rental Program
1 River Rd.
Schenectady, NY 12345

Hickok Electrical Instruments
10514 Dupont Av.
Cleveland, OH 44108

Hioki New York Corp.
46-16 235th St.
Douglaston, NY 11363

Hipotronics, Inc.
Route 22
Brewster, NY 10509

Martindale Electric Co.
1375 Hird Av.
Cleveland, OH 44107

Pacer Industries Inc.
704 E. Grand Av.
Chippewa Falls, WI 54729

Peschel Instruments Inc.
1412 Viscaya Pkwy.
Cape Coral, FL 33904

Reliant Heating & Controls Inc.
3433 Edward Av.
Santa Clara, CA 95050

Ricca-Reddington Instruments, Inc.
1400G NW 65th Av.
Plantation, FL 33313

Ross Engineering Corp.
559 Westchester Dr.
Campbell, CA 95008

Simpson Electric Co.
Division of American Gage and
   Machine Co.
853 Dundee Av.
Elgin, IL 60120

(A.W.) Sperry Instruments Inc.
245 Marcus Bl.
Hauppauge, NY 11787

(H.H.) Sticht Co., Inc.
27 Park Pl.
New York, NY 10007

TIF Instruments Inc.
3661 NW 74th St.
Miami, FL 33147

Triplett Corporation
Bluffton, OH 45817

(The) Von Corp.
P.O. Box 3566G
Birmingham, AL 35211

Weston Instruments
A Division of Sangamo Weston Inc.
614 Frelinghuysen Av.
Newark, NJ 07114

(Daniel) Woodhead Co.
A Division of Daniel Woodhead Inc.
3411 Woodhead Dr.
Northbrook, IL 60062

Yokogawa Corp. of America
5 Westchester Plaza
Elmsford, NY 10523

• **Meters, Phase Angle**
Beckwith Electric Co., Inc.
11811 62nd St.
North Largo, FL 33543

Time Mark Corp.
P.O. Box 15127
Tulsa, OK 74115

• **Meters, Relative Humidity**
Abbeon Cal Inc.
123 Gray Av.
Santa Barbara, CA 93101

Beckman Instruments Inc.
Division of Cedar Grove Operations
89 Commerce Rd.
Cedar Grove, NJ 07009

Epic Inc.
150 Nassau St.
New York, NY 10038

H-B Instrument Co.
4314 N. American St.
Philadelphia, PA 19140

• **Ohmmeters**
AEMC Corporation
No. Amer. Dist. Chauvin Arnoux
  Prod.
729 Boylston St.
Boston, MA 02116

Amprobe Instrument
Division of Core Industries Inc.
630 Merrick Rd.
Lynbrook, NY 11563

Anderson Power Products Inc.
145 Newton St.
Boston, MA 02135

Associated Research Inc.
8221 N. Kimball Av.
Skokie, IL 60076

Beckman Instruments Inc.
Division of Cedar Grove Operations
89 Commerce Rd.
Cedar Grove, NJ 07009

(James G.) Biddle Co.
Township Line & Jolly Rds.
Plymouth Meeting, PA 19462

Burnworth Tester Co.
815 Pomona Av.
El Cerrito, CA 94530

Ecos Electronics Corporation
205 W. Harrison St.
Oak Park, IL 60304

Etcon Corporation
12243 S. 71st Av.
Palos Heights, IL 60463

General Electric Co.
Instrument Products Operation
40 Federal St.
Lynn, MA 01910

Hickok Electrical Instruments
10514 Dupont Av.
Cleveland, OH 44108

Martindale Electric Co.
1375 Hird Av.
Cleveland, OH 44107

Non-Linear Systems Inc.
533 Stevens Av.
Solana Beach, CA 92075

Ross Engineering Corporation
559 Westchester Dr.
Campbell, CA 95008

Sangamo Weston, Inc.
Schlumberger Division
P.O. Box 3347
Springfield, IL 62714

Snap-On Tools Corp.
2801 80th St.
Kenosha, WI 53140

(A.W.) Sperry Instruments Inc.
245 Marcus Bl.
Hauppauge, NY 11787

(H.H.) Sticht Co., Inc.
27 Park Pl.
New York, NY 10007

TIF Instruments Inc.
3661 NW 74th St.
Miami, FL 33147

Triplett Corp.
Bluffton, OH 45817

Weston Instruments
A Division of Sangamo Weston Inc.
614 Frelinghuysen Av.
Newark, NJ 07114

(Daniel) Woodhead Co.
Division of Daniel Woodhead Inc.
3411 Woodhead Dr.
Northbrook, IL 60062

Yokogawa Corp. of America
5 Westchester Plaza
Elmsford, NY 10523

• Phase Sequence Indicators

AEMC Corp.
No. Amer. Dist. Chauvin Arnoux
  Prod.
729 Boylston St.
Boston, MA 02116

Amprobe Instrument
Division of Core Industries Inc.
630 Merrick Rd.
Lynbrook, NY 11563

Applied Electro Technology
2220 S. Anne St.
Santa Ana, CA 92704

Associated Research Inc.
8221 N. Kimball Av.
Skokie, IL 60076

(James G.) Biddle Co.
Township Line & Jolly Rds.
Plymouth Meeting, PA 19462

Ecos Electronics Corp.
205 W. Harrison St.
Oak Park, IL 60304

Epic Inc.
150 Nassau St.
New York, NY 10038

General Electric Co.
Instrument Rental Program
1 River Rd.
Schenectady, NY 12345

General Equipment & Mfg. Co. Inc.
3300 Fern Valley Rd.
Louisville, KY 40213

Hipotronics
Route 22
Brewster, NY 10509

Knopp Inc.
1307 66th St.
Oakland, CA 94608

Lark Electronics Inc.
390 Ft. George Sta.
New York, NY 10040

Martindale Electric Co.
1375 Hird Av.
Cleveland, OH 44107

(H.H.) Sticht Co., Inc.
27 Park Pl.
New York, NY 10007

Time Mark Corp.
P.O. Box 15127
Tulsa, OK 74115

Western Electro Mechanical
300 Broadway
Oakland, CA 94607

Westinghouse
Division of Relay-Instrument
95 Orange St., P.O. Box 606
Newark, NJ 07101

• Power Factor Meters

AEMC Corp.
No. Amer. Dist. Chauvin Arnoux
  Prod.
729 Boylston St.
Boston, MA 02116

(James G.) Biddle Co.
Township Line & Jolly Rds.
Plymouth Meeting, PA 19462

Ecos Electronics Corp.
205 W. Harrison St.
Oak Park, IL 60304

Epic Inc.
150 Nassau St.
New York, NY 10038

General Electric Co.
Instrument Products Operation
40 Federal St.
Lynn, MA 01910

General Electric Co.
Instrument Rental Program
1 River Rd.
Schenectady, NY 12345

Martindale Electric Co.
1375 Hird Av.
Cleveland, OH 44107

Square D Company
P.O. Box 6440
Clearwater, FL 33518

Westinghouse
Relay-Instrument Division
95 Orange St.
P.O. Box 606
Newark, NJ 07101

Yokogawa Corporation of America
5 Westchester Plaza
Elmsford, NY 10523

• **Recorders, Volt, Amp, Temp.**
Amprobe Instrument
Division of Core Industries Inc.
630 Merrick Rd.
Lynbrook, NY 11563

Duraline
Division of J.B. Nottingham & Co.,
  Inc.
75 Hoffman Ln.
Central Islip, NY 11722

General Electric Co.
Instrument Products Operation
40 Federal St.
Lynn, MA 01910

General Electric Co.
Instrument Rental Program
1 River Rd.
Schenectady, NY 12345

Genisco Technology Corp.
18435 Susana Rd.
Compton, CA 90221

Hastings Fiber Glass Prods. Inc.
770 S. Cook Rd.
Hastings, MI 49058

Martindale Electric Co.
1375 Hird Av.
Cleveland, OH 44107

Pacer Industries Inc.
704 E. Grand Av.
Chippewa Falls, WI 54729

Photron Instrument Co.
6516 Detroit Av.
Cleveland, OH 44102

Reliant Heating & Controls Inc.
3433 Edward Av.
Santa Clara, CA 95050

Sangamo Weston Inc.
Schlumberger Division
P.O. Box 3347
Springfield, IL 62714

Simpson Electric Co.
American Gage and Machine Co.
853 Dundee Av.
Elgin, IL 60120

(H.H.) Sticht Co., Inc.
27 Park Pl.
New York, NY 10007

Westinghouse
Relay-Instrument Division
95 Orange St.
P.O. Box 606
Newark, NJ 07101

Yokogawa Corp. of America
5 Westchester Plaza
Elmsford, NY 10523

• **Tachometers**
Abbeon Cal Inc.
123 Gray Av.
Santa Barbara, CA 93101

AEMC Corp.
No. Amer. Dist. Chauvin Arnoux
  Prod.
729 Boylston St.
Boston, MA 02116

(James G.) Biddle Co.
Township Line & Jolly Rds.
Plymouth Meeting, PA 19462

Dynalco Corp.
5200 NW 37th Av.
Ft. Lauderdale, FL 33310

Ecos Electronics Corp.
205 W. Harrison St.
Oak Park, IL 60304

Epic Inc.
150 Nassau St.
New York, NY 10038

General Electric Co.
Instrument Products Operation
40 Federal St.
Lynn, MA 01910

Martindale Electric Co.
1375 Hird Av.
Cleveland, OH 44107

Pacer Industries Inc.
704 E. Grand Av.
Chippewa Falls, WI 54729

Simpson Electric Co.
American Gage and Machine Co.
  Division
853 Dundee Av.
Elgin, IL 60120

Snap-On Tools Corp.
2801 80th St.
Kenosha, WI 53140

(H.H.) Sticht Co., Inc.
27 Park Pl.
New York, NY 10007

TIF Instruments Inc.
3661 NW 74th St.
Miami, FL 33147

Weston Instruments
A Division of Sangamo Weston Inc.
614 Frelinghuysen Av.
Newark, NJ 07114

• Temp. Measuring
Amprobe Instrument
Division of Core Industries Inc.
630 Merrick Rd.
Lynbrook, NY 11563

(James G.) Biddle Co.
Township Line & Jolly Rds.
Plymouth Meeting, PA 19462

Duraline
Division of J.B. Nottingham & Co.,
   Inc.
75 Hoffman Ln.
Central Islip, NY 11722

Ecos Electronics Corp.
205 W. Harrison St.
Oak Park, IL 60304

Fenwal Inc.
Div. of Walter Kidde & Co., Inc.
400 Main St.
Ashland, MA 01721

General Electric Co.
Instrument Products Operation
40 Federal St.
Lynn, MA 01910

Genisco Technology Corp.
18435 Susana Rd.
Compton, CA 90221

(Claud S.) Gordon Co.
5710 Kenosha St.
Richmond, IL 60071

H-B Instrument Co.
4314 N. American St.
Philadelphia, PA 19140

Hy-Cal Engineering
12105 Los Nietos Rd.
Santa Fe Springs, CA 90670

ITT Holub Industries
443 Elm St.
Sycamore, IL 60178

Mack Electric Devices Inc.
211 Glenside Av.
Wyncote, PA 19095

Mikron Instrument Co., Inc.
445 W. Main St.
Wyckoff, NJ 07481

Omron Electronics Inc.
233 S. Wacker Dr., #5300
Chicago, IL 60606

Payne Engineering Co.
Box 70
Scott Depot, WV 25560

RFL Industries Inc.
Powerville Rd.
Boonton, NJ 07005

Simpson Electric Co.
Division of American Gage and
   Machine Co.
853 Dundee Av.
Elgin, IL 60120

Techni-Tool Inc.
Apollo Rd.
Plymouth Meeting, PA 19462

TIF Instruments Inc.
3661 NW 74th St.
Miami, FL 33147

Triplett Corp.
Bluffton, OH 45817

Weston Instruments
A Division of Sangamo Weston Inc.
614 Frelinghuysen Av.
Newark, NJ 07114

• Thermometers
Abbeon Cal Inc.
123 Gray Av.
Santa Barbara, CA 93101

AEMC Corp.
No. Amer. Dist. Chauvin Arnoux
   Prod.
729 Boylston St.
Boston, MA 02116

Amprobe Instrument
Division of Core Industries Inc.
630 Merrick Rd.
Lynbrook, NY 11563

Fenwal Inc.
Division of Walter Kidde & Co., Inc.
400 Main St.
Ashland, MA 01721

(Claud S.) Gordon Co.
5710 Kenosha St.
Richmond, IL 60071

H-B Instrument Co.
4314 N. American St.
Philadelphia, PA 19140

Hy-Cal Engineering
12105 Los Nietos Rd.
Santa Fe Springs, CA 90670

Mack Electric Devices Inc.
211 Glenside Av.
Wyncote, PA 19095

Sagline Inc.
P.O. Box 351
Millwood, NY 10546

TIF Instruments Inc.
3661 NW 74th St.
Miami, FL 33147

United Electric Controls Co.
85 School St.
Watertown, MA 02172

Weston Instruments
A Division of Sangamo Weston Inc.
614 Frelinghuysen Av.
Newark, NJ 07114

• **Volt-Ammeters**
AEMC Corp.
No. Amer. Dist. Chauvin Arnoux
  Prod.
729 Boylston St.
Boston, MA 02116

Amprobe Instrument
Division of Core Industries Inc.
630 Merrick Rd.
Lynbrook, NY 11563

Associated Research Inc.
8221 N. Kimball Av.
Skokie, IL 60076

Burnworth Tester Co.
815 Pomona Av.
El Cerrito, CA 94530

Columbia Electric Mfg. Co.
4519 Hamilton Av.
Cleveland, OH 44114

Control Power Systems Inc.
18978 NE 4th Ct.
North Miami Beach, FL 33179

Ecos Electronics Corp.
205 W. Harrison St.
Oak Park, IL 60304

Epic Inc.
150 Nassau St.
New York, NY 10038

Etcon Corp.
12243 S. 71st Av.
Palos Heights, IL 60463

General Electric Co.
Instrument Products Operation
40 Federal St.
Lynn, MA 01910

General Electric Co.
Instrument Rental Program
1 River Rd.
Schenectady, NY 12345

Hickok Electrical Instruments
10514 Dupont Av.
Cleveland, OH 44108

Hioki New York Corp.
46-16 235th St.
Douglaston, NY 11363

ITT Holub Industries
443 Elm St.
Sycamore, IL 60178

Pacer Industries Inc.
704 E. Grand Av.
Chippewa Falls, WI 54729

Sangamo Weston Inc.
Schlumberger Division
P.O. Box 3347
Springfield, IL 62714

Simpson Electric Co.
American Gage and Machine Co.
  Division
853 Dundee Av.
Elgin, IL 60120

Snap-On Tools Corp.
2801 80th St.
Kenosha, WI 53140

(A.W.) Sperry Instruments Inc.
245 Marcus Bl.
Hauppauge, NY 11787

(H.H.) Sticht Co., Inc.
27 Park Pl.
New York, NY 10007

TIF Instruments Inc.
3661 NW 74th St.
Miami, FL 33147

Triplett Corp.
Bluffton, OH 45817

Viz Mfg. Co.
335 E. Price St.
Philadelphia, PA 19144

Western Electro Mechanical
300 Broadway
Oakland, CA 94607

Weston Instruments
A Division of Sangamo Weston Inc.
614 Frelinghuysen Av.
Newark, NJ 07114

Yokogawa Corp. of America
5 Westchester Plaza
Elmsford, NY 10523

• Voltmeters
AEMC Corp.
No. Amer. Dist. Chauvin Arnoux
    Prod.
729 Boylston St.
Boston, MA 02116

Amprobe Instrument
Division of Core Industries Inc.
630 Merrick Rd.
Lynbrook, NY 11563

Associated Research Inc.
8221 N. Kimball Av.
Skokie, IL 60076

Burnworth Tester Co.
815 Pomona Av.
El Cerrito, CA 94530

Columbia Electric Mfg. Co.
4519 Hamilton Av.
Cleveland, OH 44114

Control Power Systems Inc.
18978 NE 4th Ct.
North Miami Beach, FL 33179

Ecos Electronics Corp.
205 W. Harrison St.
Oak Park, IL 60304

Etcon Corp.
12243 S. 71st Av.
Palos Heights, IL 60463

Gardner Bender Inc.
6101 N. Baker Rd.
P.O. Box 23322
Milwaukee, WI 53209

General Electric Co.
Instrument Products Operation
40 Federal St.
Lynn, MA 01910

General Electric Co.
Instrument Rental Program
1 River Rd.
Schenectady, NY 12345

Hickok Electrical Instruments
10514 Dupont Av.
Cleveland, OH 44108

Hioki New York Corp.
46-16 235th St.
Douglaston, NY 11363

Hipotronics Inc.
Route 22
Brewster, NY 10509

ITT Holub Industries
443 Elm St.
Sycamore, IL 60178

Martindale Electric Co.
1375 Hird Av.
Cleveland, OH 44107

Mono-Probe Corp.
4205 Maycrest Av.
Los Angeles, CA 90032

Non-Linear Systems Inc.
533 Stevens Av.
Solana Beach, CA 92075

Pacer Industries Inc.
704 E. Grand Av.
Chippewa Falls, WI 54729

Peschel Instruments Inc.
1412 Viscaya Pkwy
Cape Coral, FL 33904

RFL Industries Inc.
Powerville Rd.
Boonton, NJ 07005

Ross Engineering Corp.
559 Westchester Dr.
Campbell, CA 95008

Rycom Instruments Inc.
9351 E. 59th St.
Raytown, MO 64133

Sangamo Weston Inc.
Schlumberger Division
P.O. Box 3347
Springfield, IL 62714

Sierra Electronic Operation
3885 Bohannon Dr.
Menlo Park, CA 94025

Simpson Electric Co.
Division of American Gage and
   Machine Co.
853 Dundee Av.
Elgin, IL 60120

Snap-On Tools Corp.
2801 80th St.
Kenosha, WI 53140

(A.W.) Sperry Instruments Inc.
245 Marcus Bl.
Hauppauge, NY 11787

Square D Company
P.O. Box 6440
Clearwater, FL 33518

(H.H.) Sticht Co., Inc.
27 Park Pl.
New York, NY 10007

TIF Instruments Inc.
3661 NW 74th St.
Miami, FL 33147

Triplett Corp.
Bluffton, OH 45817

Viz Mfg. Co.
335 E. Price St.
Philadelphia, PA 19144

(The) Von Corp.
P.O. Box 3566G
Birmingham, AL 35211

Western Electro Mechanical
300 Broadway
Oakland, CA 94607

Westinghouse
Relay-Instrument Division
95 Orange St., P.O. Box 606
Newark, NJ 07101

Weston Instruments
614 Frelinghuysen Av.
Newark, NJ 07114

Yokogawa Corp. of America
5 Westchester Plaza
Elmsford, NY 10523

• Volt-Ohm-Ammeters

AEMC Corp.
No. Amer. Dist. Chauvin Arnoux
   Prod.
729 Boylston St.
Boston, MA 02116

Amprobe Instrument
Division of Core Industries Inc.
630 Merrick Rd.
Lynbrook, NY 11563

Associated Research Inc.
8221 N. Kimball Av.
Skokie, IL 60076

(James G.) Biddle Co.
Township Line & Jolly Rds.
Plymouth Meeting, PA 19462

Control Power Systems Inc.
18978 NE 4th Ct.
North Miami Beach, FL 33179

Ecos Electronics Corp.
205 W. Harrison St.
Oak Park, IL 60304

Etcon Corp.
12243 S. 71st Av.
Palos Heights, IL 60463

General Electric Co.
Instrument Products Operation
40 Federal St.
Lynn, MA 01910

General Electric Co.
Instrument Rental Program
1 River Rd.
Schenectady, NY 12345

Hickok Electrical Instruments
10514 Dupont Av.
Cleveland, OH 44108

Hioki New York Corp.
46-16 235th St.
Douglaston, NY 11363

ITT Holub Industries
443 Elm St.
Sycamore, IL 60178

Martindale Electric Co.
1375 Hird Av.
Cleveland, OH 44107

Pacer Industries Inc.
704 E. Grand Av.
Chippewa Falls, WI 54729

RFL Industries Inc.
Powerville Rd.
Boonton, NJ 07005

Ricca-Reddington Instruments, Inc.
1400G NW 65th Av.
Plantation, FL 33313

Ross Engineering Corp.
559 Westchester Dr.
Campbell, CA 95008

Simpson Electric Co.
American Gage and Machine Co.
  Division
853 Dundee Av.
Elgin, IL 60120

Snap-On Tools Corp.
2801 80th St.
Kenosha, WI 53140

(H.H.) Sticht Co., Inc.
27 Park Pl.
New York, NY 10007

(A.W.) Sperry Instruments Inc.
245 Marcus Bl.
Hauppauge, NY 11787

Techni-Tool Inc.
Apollo Rd.
Plymouth Meeting, PA 19462

TIF Instruments Inc.
3661 NW 74th St.
Miami, FL 33147

Triplett Corp.
Bluffton, OH 45817

Viz Mfg. Co.
335 E. Price St.
Philadelphia, PA 19144

Weston Instruments
A Division of Sangamo Weston Inc.
614 Frelinghuysen Av.
Newark, NJ 07114

• **Wattmeters**
AEMC Corp.
No. Amer. Dist. Chauvin Arnoux
  Prod.
729 Boylston St.
Boston, MA 02116

Epic Inc.
150 Nassau St.
New York, NY 10038

General Electric Co.
Instrument Products Operation
40 Federal St.
Lynn, MA 01910

General Electric Co.
Instrument Rental Program
1 River Rd.
Schenectady, NY 12345

Hioki New York Corp.
46-16 235th St.
Douglaston, NY 11363

Knopp Inc.
1307 66th St.
Oakland, CA 94608

Martindale Electric Co.
1375 Hird Av.
Cleveland, OH 44107

RFL Industries Inc.
Powerville Rd.
Boonton, NJ 07005

Sangamo Weston Inc.
Schlumberger Division
P.O. Box 3347
Springfield, IL 62714

Simpson Electric Co.
American Gage and Machine Co.
853 Dundee Av.
Elgin, IL 60120

Square D Company
P.O. Box 6440
Clearwater, FL 33518

Triplett Corp.
Bluffton, OH 45817

Westinghouse
Relay Instrument Division
95 Orange St., P.O. Box 606
Newark, NJ 07101

Weston Instruments
A Division of Sangamo Weston Inc.
614 Frelinghuysen Av.
Newark, NJ 07114

Yokogawa Corp. of America
5 Westchester Plaza
Elmsford, NY 10523

Index

Abisofix Isolex, 197-98
Ac motors: troubleshooting (*see* Troubleshooting ac motors)
windings for, 216-20
Air gap, checking, 144
Air-over motors, 10
Alarm contacts, 51
Alternations, of generators, 105
Alternators, 106
elementary, 4-5
single-phase, 107-8
three-phase, 108-11
two-phase, 108-9
Ammeter, for armature testing, 162-64
A motors, 36
Amperes (amps), 12
Armature coils (*see also* Coils):
grounded, 174-75
short-circuited, 174
Armature core, 79
Armature equalizer connections, 69-70
Armatures:
form-wound, 225-26
galvanometer tests on, 166-68
growlers used for testing, 160-62,

169-70,177-78
magnetic controller, 42-44
maintenance of, 144-45
testing, 160-64, 177-78
voltmeter used to find faults in, 174-75
Armature windings (*see* Windings)
Automatic reset feature, in bimetallic overload relays, 50

Baking, 236-38
Ball bearings, 11
Bar-to-bar armature test, 174
Basket coils, 219-20
Bearings:
ball, 11
in dc motors, 77
high-speed belting and, 117
maintenance of, 140, 142-44
sleeve, 11
of split-phase motors, 17, 18, 155
troubleshooting, in ac motors, 154, 155
Belting, high-speed, selection of motors and, 117-18

265